T0214167

SpringerBriefs in Plant Science

SpringerBriefs present concise summaries of cutting-edge research and practical applications across a wide spectrum of fields. Featuring compact volumes of 50 to 125 pages, the series covers a range of content from professional to academic. Typical topics might include:

- A timely report of state-of-the art analytical techniques
- A bridge between new research results, as published in journal articles, and a contextual literature review
- A snapshot of a hot or emerging topic
- An in-depth case study or clinical example
- A presentation of core concepts that students must understand in order to make independent contributions

SpringerBriefs in Plant Sciences showcase emerging theory, original research, review material and practical application in plant genetics and genomics, agronomy, forestry, plant breeding and biotechnology, botany, and related fields, from a global author community. Briefs are characterized by fast, global electronic dissemination, standard publishing contracts, standardized manuscript preparation and formatting guidelines, and expedited production schedules.

More information about this series at http://www.springer.com/series/10080

Khawar Jabran

Role of Mulching in Pest Management and Agricultural Sustainability

 Springer

Khawar Jabran
Department of Plant Production and Technologies
Faculty of Agricultural Sciences and Technologies
Niğde Ömer Halisdemir University
Niğde, Turkey

ISSN 2192-1229   ISSN 2192-1210   (electronic)
SpringerBriefs in Plant Science
ISBN 978-3-030-22300-7   ISBN 978-3-030-22301-4   (eBook)
https://doi.org/10.1007/978-3-030-22301-4

This Springer imprint is published by the registered company Springer Nature Switzerland AG
The registered company address is: Gewerbestrasse 11, 6330 Cham, Switzerland

Preface

Agricultural researchers have been engaged to increase crop productivity in order to ensure global food security. Resources for food production are limited, and human food needs are on a continuous rise due to ever-increasing population. Hence, intensive agriculture has become a necessity, which leads to overexploitation and unsustainable use of the land, water, fertilizer, and other resources. Climate change, water shortage, and pesticide resistance are a few other important challenges that compromise the agricultural sustainability. Some innovative techniques such as mulching the soil are desired that can provide multiple benefits and contribute to natural pest suppression, and achieving agricultural sustainability. Although mulches have been used during old times, their use has regained popularity, and researchers are continuing to explore several of the benefits of using different mulching systems in the modern agricultural production. This book has discussed various aspects of employing mulches in a variety of agricultural production systems. It clearly describes the recent developments in the use of mulches in crop farming, the restrictions involved and the gained advantages. The detailed discussion based on the recent literature explicitly indicates that mulches provide several benefits in the perspective of organic crop production, dealing with the climate change and resistant pests, and attaining the agricultural sustainability. Owing to its several benefits, mulching will be a compulsory component of many of the future cropping systems.

Niğde, Turkey   Khawar Jabran
April 7, 2019

v

Acknowledgments

I greatly acknowledge the efforts and hard work of my teachers, particularly the ones who taught me English language and subject of science. Their perfect teaching in the past enabled me to write this book today. I will particularly mention Muhammad Ashraf (M.A.), Shukar-ud-Din (M.A.), Muhammad Saleem (M.A.), Professor Muhammad Yousaf Mughal, and Professor Munawar Ahmed.

Contents

Use of Mulches in Agriculture: Introduction and Concepts

Introduction

Conventional agriculture is required to be introduced with innovative techniques that may help in improving the sustainable use of resources. In the recent decades, the water is quickly becoming scarce due to climate change (that changes the rain patterns including its distribution and concentration), rise in human population, and contamination of freshwater resources by heavy metals and other pollutants. In addition to water scarcity, degradation of land resources is also on a continuous rise due to their unsustainable use. Some easy and economical options are needed that may diminish soil loss, help to conserve moisture, improve soil microbial activities, and may add nutrients to the soil.

Non-chemical pest control (including management of diseases, weeds, and insect pests) has been the aim of agricultural researchers for achieving various objectives (Jabran and Chauhan 2018a, b; Wang et al. 2018). Importantly, non-chemical pest control is inevitable for its indispensability for the organic crop production. For successful pest control in organic farming, multiple methods of non-chemical pest control are desired. Other than organic crop production, non-chemical weed control is also desired in the wake of evolution of pesticide resistance in the pests and for achieving the integrated pest control. Important of all, non-chemical pest control is desired for the sake of achieving sustainable pest control in the agricultural systems.

Mulches offer a significant role in accomplishing the goal of sustainability in modern agricultural production systems. These may help in conserving water and soil; achieving non-chemical control of insect pests, diseases, and weeds; controlling soil salinity; and adding nutrients to the soil (Kumar and Goh 1999; Zhao et al. 2014; Jabran et al. 2015a; Prosdocimi et al. 2016; Cerdà et al. 2017; Jabran and Chauhan 2018c). Mulch means any organic or fabricated material that is employed in any agricultural systems for the sake of achieving regulation of soil temperature,

K. Jabran, *Role of Mulching in Pest Management and Agricultural Sustainability*, SpringerBriefs in Plant Science, https://doi.org/10.1007/978-3-030-22301-4_1

warming or cooling of soil, weed control, and water and soil conservation (Jabran and Farooq 2007; Farooq et al. 2011a; Kasirajan and Ngouajio 2012). Mulches have also been exploited to control insect pests or diseases, adding nutrients to the soil or improving the nutrient availability and improving the activities of various biological entities in the soil.

Recently, materials such as hay, paper, cardboard, plastic, woodchips, etc. are being used as mulches for various advantages in agricultural systems. Weed suppression, disease control, water conservation, and soil temperature regulation are important among such benefits. There is plenty of literature that addresses the benefits of mulching in various agricultural systems. Most of this literature comprises of research papers, while very few review or books are available on the topic. Nevertheless, the information on utility of mulches in agricultural systems is available but mostly in patches. For instance, recently a book was edited by Orzolek (2017) that addressed the utilization of plastics (including use of plastic films as mulches) in agricultural production. The book did not include anything about the straw or other kind of mulches but narrated the production of plastics and their disposal and uses both in the vegetables and row crops (Orzolek 2017). Probably there is no single source that may fully explain the role of mulching systems in maintaining the sustainability and productivity of various agricultural systems. Hence, this book was planned to explain the implications of using various kinds of mulches in agricultural production and describe the role of mulching in agricultural sustainability. The scope of this book includes the organic or non-organic mulches being added to an agricultural field from several resources; however, this does not cover the living mulches, cover crops, or biochar. The next six chapters explain the role of mulches in controlling weeds, insect pests, and diseases, soil and water conservation, improving biological activities in the soil, regulating soil temperature, and improving the soil fertility. The introductory (current) chapter explains important concepts about mulches and discusses the types of mulches and their merits and demerits.

Types of Mulches

This is important to select the most appropriate mulch type keeping in regard the soil type, the environmental conditions, the crop, and the specific objectives for which the mulch is being used. Socio-economic situation and availability of modern agricultural technology also impact the selection or use of mulches (Erenstein 2002). Both degradable or organic (particularly the plant residue) and non-degradable/inorganic (such as plastic mulches) have been tested under different agricultural settings for their role in soil and water conservation, regulating soil temperature, enhancing the soil microbial activities, increasing organic matter, and controlling weeds. Among the mulches, black plastic (low-density polyethylene) mulch has a huge utilization in the orchards or greenhouse conditions (in high-value crops particularly) and possesses a utility in the arable agriculture as well

(Kasirajan and Ngouajio 2012; Kader et al. 2017). Both the black and transparent plastics have been used as mulches particularly in the vegetable production for a period of nearly 70 years (Lamont Jr 2017). The popularity of this mulch has witnessed an increase particularly in the water-short environments. Although black plastic mulch has been the most popular, the transparent or plastic mulches with other colors have also been evaluated for their benefits to soil and plants. High monetary and environmental costs may limit use of plastic mulches in agriculture, and degradable mulches are under-development as a substitute (Touchaleaume et al. 2016). Table 1 gives various kinds of mulches that have been tested in different kinds of agricultural systems along with their benefits.

In the wake environmental demerits of black or other colored plastic mulches, use of degradable or biodegradable mulches has been suggested (Moreno and Moreno 2008; Gao et al. 2019; Sander 2019). The degradable mulches are made from starch or cellulose (polysaccharides) obtained from crops such potato, wheat, sugarcane, or maize (or from several other crop sources) (Chiellini et al. 2004; Merino et al. 2019). These mulches are decomposable to H_2O, CO_2, remnant polymers, or other minerals in the soil environment during or toward the end of the crop growing season (Sander 2019). There are ways to stimulate degradability of these mulches if required (Thompson et al. 2019). The degradable mulches that have a moderate decomposition rate are more effective in the positive modification of soil and crop environment and the subsequent high yields (Yin et al. 2019). A higher degradation of potato or maize starch mulches has been noted under the clay soils than the sandy soils (Villena et al. 2017). Degradable mulch also declines the soil moisture evaporation but a little lower compared to the plastic mulch (Gu et al. 2017).

Fabricated mulches (such as black plastic) are expensive than the ones naturally available (e.g., straw mulch). Further, the plastic mulch does not add organic matter or other nutrients to the soil (Rao and Pathak 1998). Choice of mulch, along with other factors, should also consider the easy availability of mulch materials in the area of use. A 5-year study in China investigated the use of gravel and straw mulches for improving water contents in soil and water use efficiency and yield of alfalfa; the authors argued that they had chosen these mulches because of their easy availability in the cropping area (Jun et al. 2014).

The global straw production from various crops is more than the quantity of grains produced by these crops. This may be estimated from the quantity of grain produced and harvest index of the crops. For instance, the three top cereals (wheat, maize, and rice) produce a grain quantity of 771.7, 1134.7, and 769.7 million tons, respectively, and these crops have an average estimated harvest index of ~0.4–0.6, ~0.45–0.55, and ~0.4 for wheat, maize, and rice (Sharma and Smith 1986; Moser et al. 2006; Djaman et al. 2013; Jabran et al. 2015c; FAO 2017). This implies that quantity of straw produced by these three crops may surpass 2500 million tons. Straw of the previous crop is frequently used as a mulch in various cropping systems including those in the conservation agriculture where mulching the soil with crop residue is a principal component (Farooq et al. 2011a). However, in many of the cases, the residue or straw from the previous is either burned or removed from

Table 1 Types of mulches and their benefits or disadvantages

Type of mulch	Benefits	References
Straw mulch	Improved rice growth, grain characteristics and recovery	Jabran et al. (2015b)
	Reduced soil erosion and run-off, soil water conservation, and temperature regulation	Montenegro et al. (2013)
	Reduction in the impact of soil compaction	Siczek and Lipiec (2011)
	Increase in soil organic matter and aggregate stability and decrease in soil and water losses	García-Orenes et al. (2009)
	Increase in water use efficiency, yield, and soil moisture level in the maize fields	Wang et al. (2009)
	Decrease in soil moisture loss	Sarkar et al. (2007)
	Reduced evaporation and increased water use efficiency and cucumber yields	Kirnak and Demirtas (2006)
	Increase in the essential oil produced by a medicinal plant geranium (*Pelargonium graveolens*)	Ram et al. (2003)
	Increased crop growth rate, plant biomass, harvest index, and maize grain yield	Awal and Khan (2000)
Pale blue and white painted straw mulch	A 15% increase in potato yield	Mantheny et al. (1992)
Aluminum/black and silver/black mulch	Increase in cucumber plant growth and yield	Torres-Olivar et al. (2018)
Red colored plastic mulch	Increase in aroma compounds of strawberries	Loughrin and Kasperbauer (2002)
Coupled plastic and straw mulch	Decrease in evaporation and increase in soil moisture levels	Yin et al. (2016)

(continued)

Table 1 (continued)

Type of mulch	Benefits	References
Black-colored plastic mulch	Increase in cucumber biomass, leaf area, and yield	Torres-Olivar et al. (2018)
	Increase in gross income and net returns in the aerobic rice production	Jabran et al. (2016)
	Increase in rice leaf area, crop growth, grain length, and brown and white rice recovery	Jabran et al. (2015b)
	High soil moisture and temperature	Xiukang et al. (2015)
	An increase of 7% in wheat yields	Wang et al. (2012)
	Low evaporation and higher cucumber yield and water use efficiency	Kirnak and Demirtas (2006)
	Increased strawberry yield, fruit size, and growth	Rajbir and Ram (2005)
Degradable film	Increase in soil temperature and water at the start of growing period and a ~30% increase in yield of maize and its water use efficiency	Li et al. (2012)
Degradable film made from polycaprolactone, starch of maize, adjuvants, and grease (60%, 30%, 5%, and 5%, respectively)	A ~10% decrease in evapotranspiration and ~54% and ~38% increase in water use efficiency and seed yield of *Brassica napus* L.	Gu et al. (2017)
Water hyacinth	High harvest index, maize yield, and growth	Awal and Khan (2000)
Mulch of *Sesbania* and *Jatropha* residue	A ~10–12% increase in yield and net returns	Jat et al. (2015)
Almond shell mulch	Increase in soil enzyme activities and organic carbon	López et al. (2014)
Manure compost and bark chips mulches	Reduced accumulation of toxic compounds (polycyclic aromatic hydrocarbons) in the soil	Krzebietke et al. (2018)

the field to feed animals or to use them for other purposes (Erenstein 2002). For instance, the quantity of crop reside produced only in China counts hundreds millions of tons, and most of it is either burned as fuel or fed to animals (Gao et al. 2009). Instead of burning or removal from the field, the crop residues are desired to be used as mulching material for the subsequent benefits such as improved soil properties, increased soil organic matter status and microbial activities, and enhanced water holding capacity. If the straw is in high quantity or possesses an allelopathic potential, it is likely to provide a weed control through a physical suppression and allelopathic effect, respectively (Farooq et al. 2011a; Jabran et al. 2015; Jabran 2017a, b, c, d, e, f, g, h, i, j; Lowry and Smith 2018). Simultaneous action of both the allelopathic effect and physical suppression is also possible. Straw mulches from crops such as sunflower, rice, canola, sorghum, wheat, barley, maize, and rye have expressed an allelopathic activity against weeds (Jabran 2017a, b, c, d, e, f, g, h, i, j). Some studies showed that straw quantity of 1 or 2 t/ha was enough to provide weed control in field conditions (Xuan et al. 2005); however, this may not be enough residue to achieve a good level of weed control (Ranaivoson et al. 2018). In most of the studies, mulch quantity of 4–6 t/ha or higher has been used, and this was found to provide reasonable weed control and also had positive effects on soil environment and crop growth (Ranaivoson et al. 2018).

Compost made from farmyard wastes or the wastes from other sources (such as municipal solid waste) have also been utilized as mulch. Compost mulch not only adds nutrients to the soil but also covers the soil to reduce evaporation, increases soil moisture status, and improves crop water use efficiency and yields. Use of compost as a weed control agent has been tested, but it did not provide encouraging results (Radics and Szné Bognár 2002). In our own research work too (data not published), the compost mulch enhanced the concentration of soil nutrients and organic matter and improved its physical properties as well but did not provide an effective weed control.

The newspapers, craft papers, and cardboards are low cost and degradable and may easily be used as mulches in vegetables or other agricultural production systems. Some unique cases of mulches are also on record and may not be limited to aluminum foil, volcanic residue, and soil and gravel. Some studies indicated the use of sawdust and water hyacinth as mulch (Awal and Khan 2000).

Benefits of Mulches

Immediate (season-long) impacts of mulches may include reduced evaporation, enhanced soil moisture retentions, improved water productivity or water use efficiency, regulation of soil temperature, and suppression of some pests. There are several long-run positive effects of mulches on soils, for example, increased soil sustainability, enhanced organic matter status of soil, addition of nutrients to the soil, and improved ecosystem services. Plastic mulches may also modify the microclimate of a crop due to the color of mulch. For instance, silver- or white-colored plastic mulch could reflect more of the incident light than the red or black (Decoteau et al. 1989). This implies that colored mulches could modify the light quantity and spectral balance within the microclimate of a crop. Red-colored plastic mulch reflected more of far-red and red light than did the black mulch, and this increased the size of strawberries and their aroma (Kasperbauer et al. 2001; Loughrin and Kasperbauer 2002). Straw mulch was applied with different colors to find out a light reflection pattern most suitable to the potato plants, and the straw painted in white and pale blue increased the yield of potato by 15% (Mantheny et al. 1992). A meta-analysis from China considered 266 scientific studies published in the peer-reviewed journals and concluded that mulches increase the water use efficiency (27.6%) and crop yields (24.3%) for potato, cotton, maize, and wheat (Gao et al. 2019).

Mulches indirectly positively impact the grain quality of rice grown under the non-flooded conditions by enhancing the retention of moisture in the soil (Jabran et al. 2015b). There is a stress to grow rice under non-flooded conditions in the wake of water scarcity. The rice grown under non-flooded conditions may face a water stress, and mulches may be used to prolong moisture retention in the soil (through decreased evaporation and improved physical properties of soil). Both the plastic and straw mulches improve the water status of soil along with an improvement in the rice growth, grain characteristics (increase in brown and white rice length and rice recovery), and crop yields (Jabran et al. 2015a, b). Nevertheless, plastic mulch earned higher yield and net returns than the straw mulch (Jabran et al. 2015a, 2016). The use of plastic mulch may be more beneficial in the water-scarce environments because this reduces the water loss and makes it available for the crop plants. A comparison of this mulch under the well-watered and water-scarce environment shows that the mulch improves the crop water use (Zhou et al. 2017).

In addition to these traditional benefits, some unique benefits of using mulches in the different agricultural systems have also been noted. For example, mulches comprising of a manure compost and bark chips could reduce the accumulation of toxic substances (such as polycyclic aromatic hydrocarbons) in the soils of hazel orchard (Krzebietke et al. 2018).

Mulches of degradable or biodegradable films are advantageous over the plastic mulches because of their decomposition in the soil, leaving no hazardous residue. A degradable mulch made from a mixture of polybutyrate adipate terephthalate plus starch (15%) had a higher concentration of chlorophyll, rate of photosynthesis, and leaf area index compared to control or other treatments in the study (Sun et al. 2018).

Disadvantages of Mulches

Most of the studies indicate that mulching benefits the soil as well as the crop plants (Table 1). In addition to positive effects, some negative impacts of the mulching are also witnessed (Changrong et al. 2006; Granatstein and Mullinix 2008; Gao et al. 2019). For instance, the ill impacts of plastic mulch residues on the soil and environment are important. Plastic mulch is non-degradable, it is difficult or even impossible to collect its remains from the field, and these residues not only hinder the filed management but also damage the aesthetic value of the agricultural fields (Pande et al. 2005). Crop yields decrease if the plastic mulch residue in the field exceeds 240 kg/ha (Gao et al. 2019). Residues of plastic mulch may hinder the water movements within soil and interfere the growth of crop plants (when present in an unmanaged manner). Further, a soil-applied herbicide or a fertilizer applied to a field containing plastic mulch residues may not reach their place of action if intercepted by a plastic mulch residue. A study from China investigated the effects of plastic mulch residues on the soil properties, water flow, distribution, and infiltration in the soil (Jiang et al. 2017). The plastic residues altered the soil properties, disturbed the water velocity inside the soil, decreased water infiltration, and disturbed the uniform water distribution in the soil (Jiang et al. 2017). Plastic mulch may also cause a decrease in soil aeration (Pande et al. 2005). Moreover, high cost of plastic mulch is another of its demerits. A study compared the cost and net returns with the use of plastic and straw mulches in the non-flooded rice (Jabran et al. 2016). Compared to the no-mulch treatment, the application of straw or plastic mulches had increased the cost of water-saving rice system, but these systems also had greater gross income. However, only plastic mulch had net returns higher than no-mulch, while these were lower for straw mulch than the no-mulch (Jabran et al. 2016).

Fabrication of degradable mulches that can substitute the plastic mulches may help to avoid the environmental and soil pollution arising through the use of plastic mulches. Straw mulches may add more of nutrients and organic matter to the soil, but the plastic mulches have a stronger effect to conserve soil moisture and suppress weeds.

Negative impacts of plastic mulches on plant health have also been noted. The black plastic mulch may cause a rise in soil temperature greater than the optimally required. Air circulation and gas exchange over the soil surface covered with plastic mulch are reduced. These factors are likely to damage some plant parts and disturb some of the plant growth processes or the uptake of nutrients. Yield is damaged and nitrogen metabolism is disturbed in potato by overheating caused by black and transparent plastic mulches, while a white or white-black mulch had increased the root temperature within optimal limits and hence a higher efficiency of nitrogen use and potato tuber yield (Ruiz et al. 1999). Many of degradable mulches allow the gas exchange between the soil and immediate environment, and this is beneficial both in terms of soil temperature regulation and gas exchange (Moreno and Moreno 2008).

Some mulches may also negatively affect the microclimate of the field, lower the soil temperature or overheat the soil, and disturb the mobilization of soil nutrients. For example, mulch containing maize straw lowered the wheat yields by lowering the soil temperature (Suying et al. 2005), and oat (*Avena sativa* L.) mulch slowed the development of maize through allelopathy and reducing soil temperature (Fortin and Pierce 1991). For weed suppression, the straw mulches not only exercise a physical pressure on weeds but also cause an allelopathic inhibition of the weed seedlings or germinating weed seeds (Farooq et al. 2011b; Jabran 2017a, b). In some cases, this is likely that straw mulch can pose allelopathic damage to crops as well. Hence the direct contact of crop plants and mulches may be avoided particularly during the seedling stage of crops. Straw mulches that are of low quality, i.e., possess a high lignin content and a wide C/N ratio, are likely to impact the soil microbe and crop's nitrogen nutrition negatively. Tian et al. (1995) developed a residue quality index in order to facilitate use of the most appropriate type of straw mulch.

Impacts on Crop Yields

Generally, mulches benefit the crop plants through water conservation, suppressing weeds and other pests, improving crop nutrition, and modifying the soil environment. Subsequently, higher crop or fruit yields are observed in the mulched plots compared to the bare soil. For instance, a 3-year study from China indicated that plastic mulch application could enhance the moisture contents and temperature in the maize fields and ultimately the grain yield in a range of 10–20% (Xiukang et al. 2015).

Black plastic mulch provided better results than straw and transparent plastic mulches (an ~18% and 24% increase in yield, respectively) in strawberry cultivation and increased the size of fruits along with early setting of flowers and fruits (Rajbir and Ram 2005). A long-term (7 year) study compared the effects of straw and plastic mulch on wheat yields, and the results indicated that straw mulch caused an average 1.46% decrease in wheat yield, while plastic mulch had caused an average of ~7% increase in wheat grain yield (Wang et al. 2012). A plastic mulch with different colors on both of its surfaces has also been investigated for its benefits to regulate soil temperature, improve light reflectance patterns, and increase plant growth and hence the yields. For instance, white/black, aluminum/black, black, and silver/black plastic mulches were tested in pickling cucumber (Torres-Olivar et al. 2018). Plant photosynthesis and nutrient concentrations were not greatly impacted, but the mulches (except white/black) increased the plant biomass, leaf area, height, and yield (Torres-Olivar et al. 2018).

Conclusions

Most of the relevant research indicate that mulches provided several benefits to plants in fruit orchards, field crops, or the ones grown under greenhouse conditions. Rarely the demerits of using mulches in agricultural systems have also been reported. Plastic mulches usually have expressed a strong influence in the fields where they were applied. Various colored plastic mulches have been tested for their benefits in the agricultural systems. Black plastic mulches have been found to be the most investigated, popular, and advantageous than other mulches. Natural or organic mulches usually include straw (maize, sunflower, rice, wheat, and others), remains of woody plants (both leaf litter and woody parts), or compost. Cardboards, newspapers, sand and gravel, carpets, etc. have also been investigated as mulches.

Several of disadvantageous of mulches are also on record. For example, high cost and pollution are two important constraints that restrict the use of black plastic mulch. Overheating of soil particularly during summer months is a disadvantage of plastic mulches. Mulches usually modify the soil environment and microclimate. Straw mulch may have more of positive effects on modifying the soil environment, while plastic mulch may be more effective against weeds than the straw mulch. Allelopathic impacts of straw mulches may negatively impact the crop plants along with a weed control; this can be avoided through proper application of straw mulch in the field.

Application of mulches in the fields is a difficult and chaotic task; hence, machinery should be developed that may accomplish the process of mulch application in the field (Bégin et al. 2001). Water scarcity, degradation of natural soil resources, and desire for the non-chemical pest control may necessitate the soil covering (mulching) a compulsory ritual in the future agriculture. Soil mulching is expected to play an inevitable role in the soil and water conservation in the near future.

References

Awal, M.A. and Khan, M.A.H., 2000. Mulch induced eco-physiological growth and yield of maize. *Pakistan Journal of Biological Sciences*, *3*(1), 61–64.

Bégin, S., Dubé, S.L. and Calandriello, J., 2001. Mulching and plasticulture. In: Physical Control Methods in Plant Protection (pp. 215–223). Springer, Berlin, Heidelberg.

Cerdà, A., Rodrigo-Comino, J., Giménez-Morera, A. and Keesstra, S.D., 2017. An economic, perception and biophysical approach to the use of oat straw as mulch in Mediterranean rainfed agriculture land. *Ecological Engineering*, *108*, 162–171.

Changrong, Y., Xurong, M., Wenqing, H. and Shenghua, Z., 2006. Present situation of residue pollution of mulching plastic film and controlling measures. *Transactions of the Chinese Society of Agricultural Engineering*, *11*, 058.

Chiellini, E., Cinelli, P., Chiellini, F. and Imam, S.H., 2004. Environmentally degradable bio-based polymeric blends and composites. *Macromolecular Bioscience*, *4*(3), 218–231.

Decoteau, D.R., Kasperbauer, M.J. and Hunt, P.G., 1989. Mulch surface color affects yield of fresh-market tomatoes. *Journal of the American Society for Horticultural Science*, *114*(2), 216–219.

Djaman, K., Irmak, S., Rathje, W.R., Martin, D.L. and Eisenhauer, D.E., 2013. Maize evapotranspiration, yield production functions, biomass, grain yield, harvest index, and yield response factors under full and limited irrigation. *Transactions of the ASABE*, *56*(2), 373–393.

Erenstein, O., 2002. Crop residue mulching in tropical and semi-tropical countries: An evaluation of residue availability and other technological implications. *Soil and Tillage Research*, 67(2), 115–133.

FAO, 2017. FAOSTAT. Crops. Food and Agriculture Organization of United Nations. Rome, Italy. Available online: http://www.fao.org/faostat/en/#data/QC/visualize. Accessed: 21.03.2019.

Farooq, M, Flower, K.C., Jabran, K., Wahid, A., Siddique, K.H.M., 2011a. Crop yield and weed management in rainfed conservation agriculture. *Soil and Tillage Research* 117: 172–183.

Farooq, M., Jabran, K., Cheema, Z.A., Wahid, A., Siddique, K.H.M., 2011b. The role of allelopathy in agricultural pest management. *Pest Management Science 67*, 493–506.

Fortin, M.C. and Pierce, F.J., 1991. Timing and nature of mulch retardation of corn vegetative development. *Agronomy Journal, 83*(1), 258–263.

Gao, H., Yan, C., Liu, Q., Ding, W., Chen, B. and Li, Z., 2019. Effects of plastic mulching and plastic residue on agricultural production: A meta-analysis. *Science of the Total Environment, 651*, 484–492.

Gao, L.W., Ma, L., Zhang, W.F., Wang, F.H., Ma, W.Q., Zhang, F.S., 2009. Estimation of nutrient resource quantity of crop straw and its utilization situation in China. *Transactions of the Chinese Society of Agricultural Engineering* 25, 173–179.

García-Orenes, F., Cerdà, A., Mataix-Solera, J., Guerrero, C., Bodí, M.B., Arcenegui, V., Zornoza, R. and Sempere, J.G., 2009. Effects of agricultural management on surface soil properties and soil–water losses in eastern Spain. *Soil and Tillage Research, 106*(1), 117–123.

Granatstein, D. and Mullinix, K., 2008. Mulching options for Northwest organic and conventional orchards. *HortScience, 43*(1), 45–50.

Gu, X.B., Li, Y.N. and Du, Y.D., 2017. Biodegradable film mulching improves soil temperature, moisture and seed yield of winter oilseed rape (*Brassica napus* L.). *Soil and Tillage Research, 171*, 42–50.

Jabran, K., 2017a. Manipulation of allelopathic crops for weed control. Springer International Publishing, USA. https://doi.org/10.1007/978-3-319-53186-1

Jabran, K., 2017b. Allelopathy: Introduction and Concepts. In: Manipulation of Allelopathic Crops for Weed Control, 1st ed.; Springer Nature International Publishing: Cham, Switzerland. pp: 1–12.

Jabran, K., 2017c. Wheat Allelopathy for Weed Control. In: Manipulation of Allelopathic Crops for Weed Control, 1st ed.; Springer Nature International Publishing: Cham, Switzerland. pp: 13–20.

Jabran, K., 2017d. *Brassicaceae* Allelopathy for Weed Control. In: Manipulation of Allelopathic Crops for Weed Control, 1st ed.; Springer Nature International Publishing: Cham, Switzerland. pp: 21–28.

Jabran, K., 2017e. Maize Allelopathy for Weed Control. In: Manipulation of Allelopathic Crops for Weed Control, 1st ed.; Springer Nature International Publishing: Cham, Switzerland. pp: 29–34.

Jabran, K., 2017f. Rice Allelopathy for Weed Control. In: Manipulation of Allelopathic Crops for Weed Control, 1st ed.; Springer Nature International Publishing: Cham, Switzerland. pp: 35–48.

Jabran, K., 2017g. Rye Allelopathy for Weed Control. In: Manipulation of Allelopathic Crops for Weed Control, 1st ed.; Springer Nature International Publishing: Cham, Switzerland. pp: 49–56.

Jabran, K., 2017h. Barley Allelopathy for Weed Control. In: Manipulation of Allelopathic Crops for Weed Control, 1st ed.; Springer Nature International Publishing: Cham, Switzerland. pp: 57–64.

Jabran, K., 2017i. Sorghum Allelopathy for Weed Control. In: Manipulation of Allelopathic Crops for Weed Control, 1st ed.; Springer Nature International Publishing: Cham, Switzerland. pp: 65–76.

Jabran, K., 2017j. Sunflower Allelopathy for Weed Control. In: Manipulation of Allelopathic Crops for Weed Control, 1st ed.; Springer Nature International Publishing: Cham, Switzerland. pp: 77–86.

Jabran, K. and Farooq, M., 2007. Mulching for resource conservation. *Daily Dawn, Lahore, Pakistan*, Feb. 26, 2007.

Jabran, K., Chauhan, B.S., 2018a. Non-Chemical Weed Control. Elsevier, Academic Press, London, United Kingdom.

Jabran, K., Chauhan, B.S., 2018b. Overview and significance of non-chemical weed control. In: Jabran, K., Chauhan. B.S. (eds.) Non-Chemical Weed Control. Elsevier, Academic Press, London, United Kingdom.

Jabran, K., Chauhan, B.S., 2018c. Weed control using ground cover systems. In: Jabran, K., Chauhan. B.S. (eds.) Non-Chemical Weed Control. Elsevier, Academic Press, London, United Kingdom.

Jabran, K., Ullah, E. and Akbar, N., 2015b. Mulching improves crop growth, grain length, head rice and milling recovery of basmati rice grown in water-saving production systems. *International Journal of Agriculture and Biology. 17*, 920–928. https://doi.org/10.17957/IJAB/15.0019.

Jabran, K., Ullah, E., Hussain, M., Farooq, M., Yaseen, M., Zaman, U. and Chauhan, B.S., 2015a. Mulching improves water productivity, yield and quality of fine rice under water-saving rice production systems. *Journal of Agronomy and Crop Science 201*, 389–400. https://doi.org/10.1111/jac.12099.

Jabran, K., Mahajan, G., Surindar, V., Chauhan, B.S., 2015. Allelopathy for weed control in agricultural systems. *Crop Protection 72*, 57–65. https://doi.org/10.1016/j.cropro.2015.03.004.

Jabran, K., Hussain, M., Fahad, S., Farooq, M., Bajwa, A.A., Alharrby, H., Nasim, W., 2016. Economic assessment of different mulches in conventional and water-saving rice production systems. *Environmental Science and Pollution Research* 23:9156–9163. https://doi.org/10.1007/s11356-016-6162-y.

Jabran, K., Ullah, E., Hussain, M., Farooq, M., Haider, N. and Chauhan, B.S., 2015c. Water saving, water productivity and yield outputs of fine-grain rice cultivars under conventional and water-saving rice production systems. *Experimental Agriculture, 51*(4), 567–581.

Jat, H.S., Singh, G., Singh, R., Choudhary, M., Jat, M.L., Gathala, M.K. and Sharma, D.K., 2015. Management influence on maize–wheat system performance, water productivity and soil biology. *Soil Use and Management, 31*(4), 534–543.

Jiang, X.J., Liu, W., Wang, E., Zhou, T. and Xin, P., 2017. Residual plastic mulch fragments effects on soil physical properties and water flow behavior in the Minqin Oasis, northwestern China. *Soil and Tillage Research, 166*, 100–107.

Jun, F., Yu, G., Quanjiu, W., Malhi, S.S. and Yangyang, L., 2014. Mulching effects on water storage in soil and its depletion by alfalfa in the Loess Plateau of northwestern China. *Agricultural Water Management, 138*, 10–16.

Kader, M.A., Senge, M., Mojid, M.A. and Ito, K., 2017. Recent advances in mulching materials and methods for modifying soil environment. *Soil and Tillage Research, 168*, 155-166.

Kasirajan, S. and Ngouajio, M., 2012. Polyethylene and biodegradable mulches for agricultural applications: a review. *Agronomy for Sustainable Development, 32*(2), 501–529.

Kasperbauer, M.J., Loughrin, J.H. and Wang, S.Y., 2001. Light reflected from red mulch to ripening strawberries affects aroma, sugar and organic acid concentrations. *Photochemistry and Photobiology, 74*(1), 103–107.

Kirnak, H. and Demirtas, M.N., 2006. Effects of different irrigation regimes and mulches on yield and macronutrition levels of drip-irrigated cucumber under open field conditions. *Journal of Plant Nutrition, 29*(9), 1675–1690.

Krzebietke, S.J., Wierzbowska, J., Żarczyński, P.J., Sienkiewicz, S., Bosiacki, M., Markuszewski, B., Nogalska, A. and Mackiewicz-Walec, E., 2018. Content of PAHs in soil of a hazel orchard depending on the method of weed control. *Environmental monitoring and assessment, 190*(7), 422.

Kumar, K. and Goh, K.M., 1999. Crop residues and management practices: effects on soil quality, soil nitrogen dynamics, crop yield, and nitrogen recovery. *Advances in Agronomy. 68*, 197–319.

Lamont Jr, W.J., 2017. Plastic mulches for the production of vegetable crops. In *A Guide to the Manufacture, Performance, and Potential of Plastics in Agriculture* (pp. 45–60). Elsevier.

Li, R., Hou, X., Jia, Z., Han, Q. and Yang, B., 2012. Effects of rainfall harvesting and mulching technologies on soil water, temperature, and maize yield in Loess Plateau region of China. *Soil Research, 50*(2), 105-113.

López, R., Burgos, P., Hermoso, J.M., Hormaza, J.I. and González-Fernández, J.J., 2014. Long term changes in soil properties and enzyme activities after almond shell mulching in avocado organic production. Soil and Tillage Research, *143*, 155–163.

Loughrin, J.H. and Kasperbauer, M.J., 2002. Aroma of fresh strawberries is enhanced by ripening over red versus black mulch. *Journal of Agricultural and Food Chemistry*, *50*(1), 161–165.

Lowry, C.J. and Smith, R.G., 2018. Weed Control Through Crop Plant Manipulations. In *Non-Chemical Weed Control* (pp. 73-96). Academic Press, Elsevier USA.

Mantheny, T.A., Hunt, P.G. and Kasperbauer, M.J., 1992. Potato tuber production in response to reflected light from different colored mulches. *Crop Science*, *32*(4), 1021–1024.

Merino, D., Gutiérrez, T.J. and Alvarez, V.A., 2019. Potential Agricultural Mulch Films Based on Native and Phosphorylated Corn Starch With and Without Surface Functionalization with Chitosan. *Journal of Polymers and the Environment*, *27*(1), 97–105.

Montenegro, A.A.A., Abrantes, J.R.C.B., de Lima, J.L.M.P., Singh, V.P., Santos, T.E.M., 2013. Impact of mulching on soil and water dynamics under intermittent simulated rainfall. Catena, *109*, 139–149.

Moreno, M.M. and Moreno, A., 2008. Effect of different biodegradable and polyethylene mulches on soil properties and production in a tomato crop. *Scientia Horticulturae*, *116*(3), 256–263.

Moser, S.B., Feil, B., Jampatong, S. and Stamp, P., 2006. Effects of pre-anthesis drought, nitrogen fertilizer rate, and variety on grain yield, yield components, and harvest index of tropical maize. *Agricultural Water Management*, *81*(1-2), 41–58.

Orzolek, M. 2017. A Guide to the Manufacture, Performance, and Potential of Plastics in Agriculture. Elsevier, Oxford, United Kingdom.

Pande, K.K., Dimri, D.C. and Kamboj, P., 2005. Effect of various mulches on growth, yield and quality attributes of apple. *Indian Journal of Horticulture*, *62*(2), 145–147.

Prosdocimi, M., Jordán, A., Tarolli, P., Keesstra, S., Novara, A. and Cerdà, A., 2016. The immediate effectiveness of barley straw mulch in reducing soil erodibility and surface runoff generation in Mediterranean vineyards. *Science of the Total Environment*, 547, 323–330.

Radics, L. and Szné Bognár, E., 2002. Comparison of different mulching methods for weed control in organic green bean and tomato. In *XXVI International Horticultural Congress: Sustainability of Horticultural Systems in the 21st Century 638*(pp. 189–196).

Rajbir, S., Ram, A., 2005. Growth, earliness and fruit yield of micro-irrigated strawberry as affected by planting time and mulching in semi-arid regions. *Indian Journal of Horticulture*, *62*, 148–151.

Ram, M., Ram, D., Roy, S.K., 2003. Influence of an organic mulching on fertilizer nitrogen use efficiency and herb and essential oil yields in geranium (*Pelargonium graveolens*). *Bioresource Technology*, *87*, 273–278.

Ranaivoson L., Naudin K., Ripoche A., Rabeharisoa L., Corbeels M., 2018. Is mulching an efficient way to control weeds? Effects of type and amount of crop residue in rainfed rice based cropping systems in Madagascar. Field Crops Research, *217*, 20–31.

Rao, V.K. and Pathak, R.K., 1998. Effect of mulches on aonla (*Emblica officinalis*) orchard in sodic soil. *Indian Journal of Horticulture*, *55*(1), 27–32.

Ruiz, J.M., Hernandez, J., Castilla, N., Romero, L., 1999. Potato performance in response to different mulches. 1. Nitrogen metabolism and yield. *Journal of Agricultural and Food Chemistry*, 47 (7), 2660–2665.

Sander, M., 2019. Biodegradation of polymeric mulch films in agricultural soils: concepts, knowledge gaps, and future research directions. *Environmental Science & Technology, 53* (5), 2304–2315

Sarkar, S., Paramanick, M., Goswami, S.B., 2007. Soil temperature, water use and yield of yellow sarson (*Brassica napus* L. var. glauca) in relation to tillage intensity and mulch management under rainfed lowland ecosystem in eastern India. *Soil and Tillage Research*, *93*, 94–101.

Sharma, R.C. and Smith, E.L., 1986. Selection for high and low harvest index in three winter wheat populations. *Crop Science*, *26*(6), 1147–1150.

Siczek, A. and Lipiec, J., 2011. Soybean nodulation and nitrogen fixation in response to soil compaction and surface straw mulching. *Soil and Tillage Research*, *114*(1), 50–56.

Sun, T., Li, G., Ning, T.Y., Zhang, Z.M., Mi, Q.H. and Lal, R., 2018. Suitability of mulching with biodegradable film to moderate soil temperature and moisture and to increase photosynthesis and yield in peanut. *Agricultural Water Management*, *208*, 214–223.

Suying, C., Xiying, Z., Dong, P. and Hongyong, S., 2005. Effects of corn straw mulching on soil temperature and soil evaporation of winter wheat field. *Transactions of The Chinese Society of Agricultural Engineering*, *10*, 038.

Thompson, A.A., Samuelson, M.B., Kadoma, I., Soto-Cantu, E., Drijber, R. and Wortman, S.E., 2019. Degradation Rate of Bio-based Agricultural Mulch is Influenced by Mulch Composition and Biostimulant Application. *Journal of Polymers and the Environment*, https://doi.org/10.1007/s10924-019-01371-9.

Tian, G., Brussaard, L. and Kang, B.T., 1995. An index for assessing the quality of plant residues and evaluating their effects on soil and crop in the (sub-) humid tropics. *Applied Soil Ecology*, *2*(1), 25–32.

Torres-Olivar, V., Ibarra-Jiménez, L., Cárdenas-Flores, A., Lira-Saldivar, R.H., Valenzuela-Soto, J.H. and Castillo-Campohermoso, M.A., 2018. Changes induced by plastic film mulches on soil temperature and their relevance in growth and fruit yield of pickling cucumber. *Acta Agriculturae Scandinavica, Section B—Soil & Plant Science*, *68*(2), 97–103.

Touchaleaume, F., Martin-Closas, L., Angellier-Coussy, H., Chevillard, A., Cesar, G., Gontard, N. and Gastaldi, E., 2016. Performance and environmental impact of biodegradable polymers as agricultural mulching films. *Chemosphere*, *144*, 433–439.

Villena, J., González, S., Moreno, C., Aceituno, P., Campos, J., Meco, R. and María Moreno, M., 2017, April. Degradation of sustainable mulch materials in two types of soil under laboratory conditions. In *EGU General Assembly Conference Abstracts* (Vol. 19, p. 1574).

Wang, S.J., Tian, X.H., Li, S. and Zhang, Y.H., 2012. Effects of long-term surface mulching and N addition on winter wheat yield and soil properties. *Plant Nutr. Fertil. Sci*, *18*, 291–299.

Wang, X., Jia, Z.K., Han, Q.F., Yang, B.P. and Nie, J.F., 2009. Effects of different straw mulching quantity on soil water and WUE in semiarid region. *Agricultural Research in the Arid Areas*, *27*(4), 196–202.

Wang, Y., Wang, Y. and Zhu, Y., 2018. What could encourage farmers to choose non-chemical pest management? Evidence from apple growers on the Loess Plateau of China. *Crop Protection*, *114*, 53–59.

Xiukang, W., Zhanbin, L. and Yingying, X., 2015. Effects of mulching and nitrogen on soil temperature, water content, nitrate-N content and maize yield in the Loess Plateau of China. *Agricultural Water Management*, *161*, 53–64.

Xuan, T.D., Shinkichi, T., Khanh, T.D. and Chung, I.M., 2005. Biological control of weeds and plant pathogens in paddy rice by exploiting plant allelopathy: an overview. *Crop Protection*, *24*(3), 197–206.

Yin, M., Li, Y., Fang, H. and Chen, P., 2019. Biodegradable mulching film with an optimum degradation rate improves soil environment and enhances maize growth. *Agricultural Water Management*, *216*, 127–137.

Yin, W., Feng, F., Zhao, C., Yu, A., Hu, F., Chai, Q., Gan, Y. and Guo, Y., 2016. Integrated double mulching practices optimizes soil temperature and improves soil water utilization in arid environments. *International Journal of Biometeorology*, *60*(9), 1423–1437.

Zhao, Y., Pang, H., Wang, J., Huo, L. and Li, Y., 2014. Effects of straw mulch and buried straw on soil moisture and salinity in relation to sunflower growth and yield. *Field Crops Research*, *161*, 16–25.

Zhou, L., Feng, H., Zhao, Y., Qi, Z., Zhang, T., He, J. and Dyck, M., 2017. Drip irrigation lateral spacing and mulching affects the wetting pattern, shoot-root regulation, and yield of maize in a sand-layered soil. *Agricultural Water Management*, *184*, 114–123.

Mulches for Weed Control

Introduction

Weeds are an unavoidable component of agricultural fields causing a significant decrease in crop productivity. In addition to decreased productivity, weeds also host the insect pests or disease pathogens, negatively impact the crop growth, strangle the field operations, and decline the quality of produce (Oerke 2006).

Herbicides have been providing a reliable weed control in most of the cropping systems since seven decades. However, recently non-chemical weed control methods are inevitable due to the issues of environmental pollution and herbicide-resistant weeds (Pannacci et al. 2017; Jabran and Chauhan 2018a, b). Mulches are important among non-chemical weed control methods. Application of mulches in agricultural fields can help to fully or partially suppress the weeds (Abouziena and Haggag 2016; Hammermeister 2016). Most of the studies report that mulches could effectively control weeds. For example, mulch of rice residue was effective in suppressing the weeds of wheat by 60–69% (Nawaz et al. 2017). Similarly, compared to mechanical weed control, straw mulch could better control the broad-leaved weeds in organically grown potato, but both the control methods were similar in the control of grass weeds (Genger et al. 2018). However, very few studies have shown that mulches were not effective in controlling weeds. For example, the studies conducted by Moore et al. (1994) indicated that mulches could not effectively control the *Amaranthus retroflexus* L. and *Chenopodium album* L. Similarly, mulching may not be a viable option for effective control of weeds due to low amount of plant residues available for application as mulch (Ranaivoson et al. 2018).

Several types of mulches can be applied to control weeds in different cropping systems. Plastic and straw mulches have been mostly investigated for their effectiveness to control weeds (Steinmetz et al. 2016; Grundy and Bond 2007). Black plastic mulch is getting popular in vegetable production systems for achieving an effective weed control along with several other benefits. Further, the mulches are

© The Author(s), under exclusive licence to Springer Nature Switzerland AG 2019

K. Jabran, *Role of Mulching in Pest Management and Agricultural Sustainability*, SpringerBriefs in Plant Science, https://doi.org/10.1007/978-3-030-22301-4_2

also important for weed control in fruit and ornamental plant nurseries. Along with being an individual technique for weed control, mulching may also play a role in integrated weed management (Swanton and Weise 1991; Jabran and Chauhan 2018a; Jabran and Chauhan 2018c). Hence, the objective of this chapter is to explain the impact of different kinds of mulches on the growth of weeds and their control. Implications of using plastic, straw, and paper mulches for weed control have been discussed in detail and other possible mulch candidate materials.

Plastic Mulch for Weed Control

Both colored and non-colored (transparent) mulches have been evaluated for suppressing weeds. However, colored mulches had a greater effectiveness against weeds. Currently, mostly black-colored plastic mulches are employed to control weeds. Like all plants, weeds require sunlight (and other resources) for their photosynthetic activity. Colored (black) plastic mulches obstruct the sunlight to reach the weeds, suppress the weeds through a physical pressure, and also restrict the germination and subsequent establishment of weed flora in an agricultural field (Ngouajio and Ernest 2004). Plastic mulches of various colors were tested for their light transmission (between 400 and 1100 nm wavelength) and weed suppression (Ngouajio and Ernest 2004). Main weeds in the experiment were *Spergula arvensis* L., *Elytrigia repens* (L.) Nevski, *Portulaca oleracea* L., *Chenopodium album* L., and *Vicia villosa* Roth. Black plastic mulch had the lowest light transmission (1%) and highest weed suppression (close to 100%), while the white plastic mulch had the highest light transmission (45%) and lowest weed control efficacy. Infrared-transmitting brown and green plastic mulches transmitted light by 26% and 42%, while a comparatively lower light transmission (17%) was noted for gray-colored plastic mulches. Considering the photosynthetically active radiation too, the black plastic mulch was the most effective among the colored or transparent mulches in reducing the light transmission and suppressing weeds (Ngouajio and Ernest 2004). Another study from Canada compared the light transmission through various types of mulches (Brault et al. 2002). A black paper and a double-layered black/white mulch had the lowest light transmission (>1% and 2%, respectively) (Brault et al. 2002). Black plastic mulch is expected to provide a complete and reliable weed control in' agricultural fields (particularly vegetable cultivation), but these may express a weaker control of sedge weeds such as *Cyperus rotundus* L. (Lament 1993). Among the black and transparent plastic mulches, transparent mulch had a lower efficacy against weeds in maize crop, while application of black plastic mulch helped to reduce the weed biomass and produce leaf area and crop biomass similar to hand weeding (Gul et al. 2009).

Hembry and Davies (1993) evaluated black polyethylene as mulch for controlling weeds in cabbage and cauliflowers and found that the mulch could provide a weed control and yield comparable to herbicide treatment. In another study, plastic mulch was tested for controlling weeds in tomato and green bean (Radics and Szné

Bognár 2002). The results indicated that the mulch could provide weed control result that was even better than herbicide application and could significantly increase the yield of both tomato and green bean (Radics and Szné Bognár 2002). A study from India used black plastic mulch for weed control in baby corn (Mahajan et al. 2007). The mulch caused a more than 60% decrease in both the weed biomass and density and also increased corn grain yield by nearly 38% (Mahajan et al. 2007).

Plastic mulch was applied in groundnut (*Arachis hypogaea* L.) to achieve weed control and other benefits (Ramakrishna et al. 2006). Important weeds in the experimental area were *Cynodon dactylon* L., *Lagascea mollis* Cav., *Galinsoga parviflora* Cav., *Eleusine indica* L., *Echinochloa colona* (L.) Link, and *Celosia argentea* L. The mulch decreased the weed score by more than 80% and weed biomass by 70% and also increased the groundnut pod yield by more than 90% along with an increase in pod weight, plant biomass, and number of pods per plant (Ramakrishna et al. 2006).

Environmental impacts, difficulties in post-season management, and cost are important concerns regarding the use of plastic mulch in the agricultural settings. Keeping in view the negative impacts of plastic mulch on the environment, degradable or biodegradable plastic (and other) mulches have also been developed. Hence, researchers suggest using degradable plastic mulches instead of the conventional ones (Prill 2017). For instance, a study from Spain compared the conventional and biodegradable plastic mulches for controlling weeds in tomato (Cirujeda et al. 2012). Both types of mulches provided equal weed control (>80%) and an increase in tomato yield (>70% of control) (Cirujeda et al. 2012). In another study too, degradable plastic mulch could control weeds such as *C. album*, *D. sanguinalis*, and *P. oleracea* in the tomato crop (Anzalone et al. 2010).

Straw Mulch for Weed Control

Field crops are usually grown to fetch their fruits (grains) for use as food, while herbage is either fed to animals or goes wasted (Erenstein 2002). Utilizing these crop residues to maintain a mulch layer in crop fields is one of the principles of conservation agriculture (Farooq et al. 2011a). A straw layer (mulch) over the soil could facilitate weed suppression along with providing other benefits (Grundy and Bond 2007; Farooq et al. 2011a, b). This weed suppression results from either the allelopathic impact of straw mulch or its physical pressure on weeds or both (Barnes and Putnam 1983; Cheema et al. 2004; Grundy and Bond 2007; Jabran et al. 2015; Jabran and Chauhan 2018b). Straws of several allelopathic crops have been used as mulch for the purpose of suppressing weeds in the field crops (Jabran 2017a, b, c, d, e, f, g, h, i, j).

Residue from the previous crop can be used as mulch, or occasionally the crop remains from other fields are acquired to be used for weed control in other fields. Another way is to grow cover crops and then kill these to use as mulch for weed control (Tursun et al. 2018). For example, oat (*Avena sativa* L.), hairy vetch (*Vicia villosa* Roth.), and subclover (*Trifolium subterraneum* L.) were grown as cover crop and then mowed in order to be used as mulch for weed control in tomato (*Lycopersicon esculentum* Mill.) (Campiglia et al. 2010a, b). The most important weeds in the experimental field were *Poa annua* L., *P. oleracea, Stellaria media* (L.) Vill., *Amaranthus retroflexus* L., and *D. sanguinalis*. Mulching could reduce weed infestation by 40–72%. Mulch application in strips was found better than the uniform scattering. Highest weed suppression was noted by oat; however, it also caused negative impacts on tomato yield, while other mulches in the study had improved the tomato yield (Campiglia et al. 2010a, b). Similarly, cover crops including *Fagopyrum esculentum, Phacelia tanacetifolia*, and *V. villosa* were converted to mulches and found effective in controlling weeds in apricot (Tursun et al. 2018). A study from the USA used cowpea [*Vigna unguiculata* (L.) Walp.] mulch for controlling weeds in pepper (*Capsicum annuum* L.) (Hutchinson and McGiffen 2000). Cowpea was grown to gain sufficient biomass and killed into a mulch before sowing of pepper. Major weeds in the experiment were *Chenopodium album, Cyperus rotundus, Amaranthus retroflexus*, and *Portulaca oleracea*. Cowpea mulch caused a high decrease in density (80–90%) and biomass (67–90%) of weeds over control, almost doubled the pepper plant biomass, and increased the fruit yield (Hutchinson and McGiffen 2000).

Straw mulch from wheat (*Triticum aestivum* L.) was evaluated for controlling weeds in tomato; however, this did not result in excellent weed control (Monks et al. 1997). Another study also tested the maize residue and straws of barley and rice (each 1 kg/m^2) for weed control in tomato (Anzalone et al. 2010). In the study, the major weeds were *Chenopodium album* L., *Digitaria sanguinalis* (L.) Scop., *Cyperus rotundus* L., and *Portulaca oleracea* L. The weeds were suppressed by the mulches except *C. rotundus*, and rice straw was the best among the mulches (Anzalone et al. 2010). In a similar study, rye and buckwheat mulches (0, 10 and 20 t/ha) were applied for controlling weeds in tomato and broccoli (Kosterna 2014). *Elymus repens* (L.) Gould., *Viola arvensis* L., *Chenopodium album* L., and *Echinochloa crus-galli* (L.) P. Beauv. were the major weeds in the experimental field. Both mulches significantly decreased the biomass and density of weeds in both the vegetables and improved their yield. A mulch rate of 20 t/ha had provided better weed control; however, 10 t/ha gave a higher broccoli and tomato yield (Kosterna 2014).

Wheat straw mulch had promising results for weed control in maize crop and could effectively suppress the grass weeds in a continuous no-till system (Crutchfield et al. 1986; Mulvaney et al. 2011). A study from Lithuania indicated that the wheat straw, wood chips, and peat mulches significantly decreased weed germination in *Phaseolus vulgaris* L. and *Allium cepa* L. (Jodaugienė et al. 2006). A greenhouse study in Mexico evaluated the leaves of dried plants as mulches for weed control in tomato (Caamal-Maldonado et al. 2001). The mulches were mixed in the soil to assess their impact on weeds and other pathogens. Wild tamarind (*Lysiloma latisiliquum* (L.) Benth.) and jackbean [*Canavalia ensiformis* (L.) DC.) caused the highest suppression of weeds, but these also had an extremely negative impact on tomato growth (Caamal-Maldonado et al. 2001). More than 40% decrease in weed biomass and count and a 28% increase in grain yield were recorded after application of rice straw mulch (at 4 t/ha) in baby corn (Mahajan et al. 2007). Rice straw mulch was evaluated for its weed suppressing ability in groundnut (Ramakrishna et al. 2006). The mulch decreased the weed score by ~59% and increased the pod yield by ~40–69% (Ramakrishna et al. 2006).

Three rates (3.5, 7 and 10 t/ha) of sorghum (*Sorghum bicolor* L.) were applied as mulch (scattered between the rows) for controlling weeds in cotton (*Gossypium arboreum* L.) (Cheema et al. 2000). *Cynodon dactylon* (L.) Pers., *Trianthema portulacastrum* L., *Convolvulus arvensis* L., and *Cyperus rotundus* L. were the important weeds infesting the research field. The sorghum mulch (all rates) could reduce the total weed biomass by nearly 50%. The first two rates of sorghum mulch had increased seed cotton yield by 53–69%, while the highest mulch rate had doubled the seed cotton yield compared to the control treatment (Cheema et al. 2000). Mulches of mushroom compost and barley straw were tested for weed control in *Beta vulgaris* ssp. *rapaceae atrorubra* Krass (Yordanova and Gerasimova 2016). Major weeds in the study were *Amaranthus retroflexus, Digitaria sanguinalis, Echinochloa crus-galli,* and *Galinsoga parviflora.* The mulches provided a significant suppression of weeds along with 8–10 times increase in crop yield over the control.

Agricultural Wastes and Manures for Weed Control

Composts and other organic manures are usually added to soil for the purpose of improving soil quality and other soil properties. Nevertheless, some role of compost mulching in suppressing weeds has been noted previously. For instance, use of compost was helpful in suppressing weeds in the vineyard (Pinamonti 1998). In apple orchards too, the compost mulch made from mushroom waste gave a good degree of weed control but also decreased the apple yields (Saoir and Mansfield 1998). Mulch of poultry manure compost too had effectively suppressed weeds in apple orchards, and this weed control was comparable to that obtained by glyphosate application in the same orchard (Brown and Tworkoski 2004). Similarly, compost of solid municipal waste also provided some encouraging results for weed control in vegetables (Roe et al. 1993).

Contrasting results are also reported where various types of compost are not effective in controlling weeds. For example, Radics and Szné Bognár (2002) evaluated compost mulch for controlling weeds in green bean and tomato, but the mulch did not provide satisfactory results for weed control or increase in the yield over the control. Results from our own work also indicated that application of compost as mulch was helpful in improving the soil properties such as water holding capacity, but it did not help in effective suppression of weeds in the direct seeded rice (data not published).

Paper Mulch for Weed Control

Although use of paper mulch for weed control has been tested frequently, the number of studies on paper mulch may be lower than the plastic mulch. Nevertheless, paper mulch possesses an advantage over the plastic mulch because paper mulch is decomposable/degradable and does not leave residue in the field. Hence, paper mulch may provide an opportunity to avoid the environmental pollution which is caused if the plastic mulch is applied in fields for weed control. Further, by end of season, it may be more convenient to manage paper mulch in field compared to the plastic mulch.

Previous literature indicates that various types of paper mulches have been tested for weed control particularly in the vegetable crops. These are not limited to craft paper, newspaper, and cardboard. Black paper was used as a mulch, and it provided almost complete control of both monocotyledonous and dicotyledonous weeds although a few dicotyledonous weeds were germinated, but these did not grow well (Brault et al. 2002). A study from the USA tested the sliced and chopped newspapers for controlling weeds in tomato (Monks et al. 1997). The newspaper (either sliced or chopped) with a depth of 7.6 cm could control the weeds by 90% or similar to chemical control. Mulch of sliced newspaper gave lower tomato yields than the chopped newspaper mulch (Monks et al. 1997). Another study from Spain indicated that craft paper mulch could not only control the broad-leaved weeds but could also better suppress the *C. rotundus* than the straw or biodegradable plastic mulch (Anzalone et al. 2010).

Paper mulch could provide better weed control and higher yield in green bean and tomato than the hoeing and herbicide application (Radics and Szné Bognár 2002). The efficacy of paper mulch was impacted by the soil moisture conditions, and it had a higher weed suppression under dry soil conditions (Radics and Szné Bognár 2002). A study from Spain compared paper mulch with plastic and straw mulches for weed control in tomato (Cirujeda et al. 2012). The results showed that paper mulch was equally effective in suppressing the weeds as do the plastic mulches. Paper mulch could also increase the tomato yield to almost double of the control treatment (Cirujeda et al. 2012).

Other Mulches for Weed Control

Black plastic, straw, and paper are the most important among mulches; there may be several other types of mulches that could be tested or used to suppress weeds in various agricultural settings (Splawski et al. 2016; Bartley et al. 2017). In this section, a few miscellaneous types of mulches are described. For example, Billeaud and Zajicek (1989) evaluated oak hardwood and pine bark as mulches (5, 10, and 15 cm depth) for weed control in *Ligustrum japonicum*. Both the mulches significantly decreased the weeds and also improved the soil properties. In another similar study, pine bark, pine needles, and hardwood were effective in suppressing the weeds by nearly 50% in the landscape areas (Skroch et al. 1992).

Small cuttings of trees such as lespedeza (*Lespedeza cuneata* (Dum.Cours.) G.Don) and mimosa (*Albizia julibrissin* Durazz.) were applied as mulch in the continuous no-till system sown with collard (*Brassica oleracea* L.) (Mulvaney et al. 2011). These mulches were most effective against sedge and broad-leaved weeds but also satisfactorily controlled the grasses toward the end of a 3-year period (Mulvaney et al. 2011). Leaves from trees could also be applied as mulch for controlling weeds in vegetables, landscape settings, or even field crops. For instance, a study evaluated leaves from three trees (*Gliricidia sepium, Leucaena leucocephala, Flemingia macrophylla*) as mulches to suppress weeds (Budelman 1988). Mulch containing leaves of *F. macrophylla* was found the most promising in controlling weeds for long durations (Budelman 1988). In a unique work, black polypropylene gave effective weed control and high yields in cauliflowers and cabbage, although this mulch caused slight decrease in yield only in one of the harvests (Hembry and Davies 1993).

Conclusions

The discussion in this chapter establishes that mulches have the potential to suppress the weeds. This provides a good possibility to use mulches as a part of integrated weed control. The potential of mulches to control weeds has not been fully explored yet, and more research and research funding are required on this aspect. Nevertheless, sometimes the mulches may be non-sustainable or cause phytotoxicity, but more of studies report the positive effects of mulches on plant growth and the negative effects on weeds. Negative impacts of mulches on crop plants may be due to careless or inappropriate use of fertilizer or a change in chemical processes of fertilizers under mulches—this is required to be studied further (Hammermeister 2016). Straw of weeds has also been occasionally investigated for use as mulch to control weeds (Khalid et al. 2018); however, it does not seem a suitable idea to use weed straw for weed control due to likeliness of addition of weed-seeds to soil and constraints in the availability of weed straw.

Plastic, straw, and paper are the three most important mulches for weed control. Plastic mulch provides a higher weed control than the other two mulch types, but paper and straw mulches may have an advantage over the plastic mulch in the environmental security perspective. Moreover, weather conditions can impact the effectiveness of mulches; many of the mulches may provide better weed control under dry soil condition (Radics and Szné Bognár 2002). Importantly, if tree cuttings/remains are used as mulch, ensure that their residue has no tree seeds to avoid them from being weed in the applied area.

References

Abouziena, H.F. and Haggag, W.M., 2016. Weed control in clean agriculture: a review. *Planta Daninha*, *34*(2), 377–392.

Anzalone, A., Cirujeda, A., Aibar, J., Pardo, G. and Zaragoza, C., 2010. Effect of biodegradable mulch materials on weed control in processing tomatoes. *Weed Technology*, *24*(3), 369–377.

Barnes, J. P., and A. R. Putnam., 1983. Rye residues contribute weed suppression in no-tillage cropping systems." *Journal of Chemical Ecology* 9: 1045–1057.

Bartley, P.C., Wehtje, G.R., Murphy, A.M., Foshee, W.G. and Gilliam, C.H., 2017. Mulch type and depth influences control of three major weed species in nursery container production. *HortTechnology*, *27*(4), 465–471.

Billeaud, L.A. and Zajicek, J.M., 1989. Influence of mulches on weed control, soil pH, soil nitrogen content, and growth of Ligustrum japonicum. *Journal of Environmental Horticulture*, *7*(4), 155–157.

Brault, D., Stewart, K.A. and Jenni, S., 2002. Optical properties of paper and polyethylene mulches used for weed control in lettuce. *HortScience*, *37*(1), 87–91.

Brown, M.W. and Tworkoski, T., 2004. Pest management benefits of compost mulch in apple orchards. *Agriculture, Ecosystems & Environment*, *103*(3), 465–472.

Budelman, A., 1988. The performance of the leaf mulches of *Leucaena leucocephala*, *Flemingia macrophylla* and *Gliricidia sepium* in weed control. *Agroforestry Systems*, *6*(1-3), 137–145.

Caamal-Maldonado, J.A., Jiménez-Osornio, J.J., Torres-Barragán, A. and Anaya, A.L., 2001. The use of allelopathic legume cover and mulch species for weed control in cropping systems. *Agronomy Journal*, *93*(1), 27–36.

Campiglia, E., Caporali, F., Radicetti, E. and Mancinelli, R., 2010a. Hairy vetch (*Vicia villosa* Roth.) cover crop residue management for improving weed control and yield in no-tillage tomato (*Lycopersicon esculentum* Mill.) production. *European Journal of Agronomy*, *33*(2), 94–102.

Campiglia, E., Mancinelli, R., Radicetti, E. and Caporali, F., 2010b. Effect of cover crops and mulches on weed control and nitrogen fertilization in tomato (*Lycopersicon esculentum* Mill.). *Crop Protection*, *29*(4), 354–363.

Cheema Z.A., Asim, M., Khaliq, A. 2000. Sorghum allelopathy for weed control in cotton (*Gossypium arboreum* L.). *International Journal of Agriculture and Biology* 2: 37–41.

Cheema, Z.A., Khaliq, A. and Saeed, S., 2004. Weed control in maize (*Zea mays* L.) through sorghum allelopathy. *Journal of Sustainable Agriculture*, *23*(4), 73–86.

Cirujeda, A., Aibar, J., Anzalone, Á., Martín-Closas, L., Meco, R., Moreno, M.M., Pardo, A., Pelacho, A.M., Rojo, F., Royo-Esnal, A. and Suso, M.L., 2012. Biodegradable mulch instead of polyethylene for weed control of processing tomato production. *Agronomy for Sustainable Development*, *32*(4), 889–897.

Crutchfield, D.A., Wicks, G.A. and Burnside, O.C., 1986. Effect of winter wheat (Triticum aestivum) straw mulch level on weed control. *Weed Science*, 110–114.

Erenstein, O., 2002. Crop residue mulching in tropical and semi-tropical countries: An evaluation of residue availability and other technological implications. *Soil and Tillage Research*, *67*(2), 115–133.

Farooq, M., Flower, K.C., Jabran, K., Wahid, A. and Siddique, K.H.M., 2011a. Crop yield and weed management in rainfed conservation agriculture. *Soil and Tillage Research*, *117*, 172–183.

Farooq, M., Jabran, K., Cheema, Z.A., Wahid, A. and Siddique, K.H.M., 2011b. The role of allelopathy in agricultural pest management. *Pest Management Science*, *67*(5), 493–506.

Genger, R.K., Rouse, D.I. and Charkowski, A.O., 2018. Straw mulch increases potato yield and suppresses weeds in an organic production system. Biological Agriculture & Horticulture, *34*(1), 53–69.

Grundy, A.C. and Bond, B., 2007. Use of non-living mulches for weed control. *Non-Chemical Weed Management*, pp.135–153.

Gul, B., Marwat, K.B., Hassan, G., Khan, A., Hashim, S. and Khan, I.A., 2009. Impact of tillage, plant population and mulches on biological yield of maize. *Pakistan Journal of Botany*, *41*(5), 243–2249.

Hammermeister, A.M., 2016. Organic weed management in perennial fruits. *Scientia Horticulturae*, *208*, 28–42.

Hembry, J.K. and Davies, J.S., 1993. Using mulches for weed control and preventing leaching of nitrogen fertiliser. In *VII International Symposium on Timing Field Production of Vegetables 371* (pp. 311–316).

Hutchinson, C.M. and McGiffen, M.E., 2000. Cowpea cover crop mulch for weed control in desert pepper production. *HortScience*, *35*(2), 196–198.

Jabran, K., 2017a. Manipulation of Allelopathic Crops for Weed Control, 1st ed.; Springer Nature International Publishing: Cham, Switzerland.

Jabran, K., 2017b. Allelopathy: Introduction and Concepts. In: Manipulation of Allelopathic Crops for Weed Control, 1st ed.; Springer Nature International Publishing: Cham, Switzerland. pp: 1–12.

Jabran, K., 2017c. Wheat Allelopathy for Weed Control. In: Manipulation of Allelopathic Crops for Weed Control, 1st ed.; Springer Nature International Publishing: Cham, Switzerland. pp: 13–20.

Jabran, K., 2017d. *Brassicaceae* Allelopathy for Weed Control. In: Manipulation of Allelopathic Crops for Weed Control, 1st ed.; Springer Nature International Publishing: Cham, Switzerland. pp: 21–28.

Jabran, K., 2017e. Maize Allelopathy for Weed Control. In: Manipulation of Allelopathic Crops for Weed Control, 1st ed.; Springer Nature International Publishing: Cham, Switzerland. pp: 29–34.

Jabran, K., 2017f. Rice Allelopathy for Weed Control. In: Manipulation of Allelopathic Crops for Weed Control, 1st ed.; Springer Nature International Publishing: Cham, Switzerland. pp: 35–48.

Jabran, K., 2017g. Rye Allelopathy for Weed Control. In: Manipulation of Allelopathic Crops for Weed Control, 1st ed.; Springer Nature International Publishing: Cham, Switzerland. pp: 49–56.

Jabran, K., 2017h. Barley Allelopathy for Weed Control. In: Manipulation of Allelopathic Crops for Weed Control, 1st ed.; Springer Nature International Publishing: Cham, Switzerland. pp: 57–64.

Jabran, K., 2017i. Sorghum Allelopathy for Weed Control. In: Manipulation of Allelopathic Crops for Weed Control, 1st ed.; Springer Nature International Publishing: Cham, Switzerland. pp: 65–76.

Jabran, K., 2017j. Sunflower Allelopathy for Weed Control. In: Manipulation of Allelopathic Crops for Weed Control, 1st ed.; Springer Nature International Publishing: Cham, Switzerland. pp: 77–86.

Jabran, K., Chauhan, B.S., 2018a. Overview and significance of non-chemical weed control. In: Jabran, K., Chauhan. B.S. (eds.) Non-Chemical Weed Control. Elsevier, Academic Press, London, United Kingdom.

Jabran, K., Chauhan, B.S., 2018b. Weed Control Using Ground Cover Systems. In: Jabran, K., Chauhan. B.S. (eds.) Non-Chemical Weed Control. Elsevier, Academic Press, London, United Kingdom.

Jabran, K., Chauhan, B.S., 2018c. Non-Chemical Weed Control. Elsevier, Academic Press, London, United Kingdom.

Jabran, K., Mahajan, G., Sardana, V. and Chauhan, B.S., 2015. Allelopathy for weed control in agricultural systems. *Crop Protection*, *72*, 57–65.

Jodaugienė, D., Pupalienė, R., Urbonienė, M., Pranckietis, V. and Pranckietienė, I., 2006. The impact of different types of organic mulches on weed emergence. *Agronomy Research*, *4*, 197–201.

Khalid, S., Shehzad, M., Zahoor, F., Mubeen, K., Ahmad, A. and Ali, E., 2018. *Parthenium hysterophorus* herbage mulching: a potential source of weeds control in soybean (*Glycine max*). *Planta Daninha, 36*.

Kosterna, E., 2014. The effect of different types of straw mulches on weed-control in vegetables cultivation. *Journal of Ecological Engineering, 15*(4): 109–117.

Lament, W.J., 1993. Plastic mulches for the production of vegetable crops. *HortTechnology, 3*(1), 35–39.

Mahajan, G., Sharda, R., Kumar, A. and Singh, K.G., 2007. Effect of plastic mulch on economizing irrigation water and weed control in baby corn sown by different methods. *African Journal of Agricultural Research, 2*(1), 19–26.

Monks, C.D., Monks, D.W., Basden, T., Selders, A., Poland, S. and Rayburn, E., 1997. Soil temperature, soil moisture, weed control, and tomato (*Lycopersicon esculentum*) response to mulching. *Weed Technology*, 561–566.

Moore, M.J., Gillespie, T.J. and Swanton, C.J., 1994. Effect of cover crop mulches on weed emergence, weed biomass, and soybean (*Glycine max*) development. *Weed Technology*, 512–518.

Mulvaney, M.J., Price, A.J. and Wood, C.W., 2011. Cover crop residue and organic mulches provide weed control during limited-input no-till collard production. *Journal of Sustainable Agriculture, 35*(3), 312–328.

Nawaz, A., Farooq, M., Lal, R., Rehman, A., Hussain, T. and Nadeem, A., 2017. Influence of sesbania brown manuring and rice residue mulch on soil health, weeds and system productivity of conservation rice–wheat systems. *Land Degradation & Development, 28*(3), 1078–1090.

Ngouajio, M. and Ernest, J., 2004. Light transmission through colored polyethylene mulches affects weed populations. *HortScience, 39*(6), 1302–1304.

Oerke, E.C., 2006. Crop losses to pests. *The Journal of Agricultural Science, 144*(1), 31–43.

Pannacci, E., Lattanzi, B. and Tei, F., 2017. Non-chemical weed management strategies in minor crops: A review. *Crop Protection, 96*, 44–58.

Pinamonti, F., 1998. Compost mulch effects on soil fertility, nutritional status and performance of grapevine. *Nutrient Cycling in Agroecosystems, 51*(3), 239–248.

Prill, S., 2017. Biodegradable Plant Mulch Device. U.S. Patent Application 15/178,116.

Radics, L. and Szné Bognár, E., 2002. Comparison of different mulching methods for weed control in organic green bean and tomato. In *XXVI International Horticultural Congress: Sustainability of Horticultural Systems in the 21st Century 638*(pp. 189–196).

Ramakrishna, A., Tam, H.M., Wani, S.P. and Long, T.D., 2006. Effect of mulch on soil temperature, moisture, weed infestation and yield of groundnut in northern Vietnam. *Field Crops Research, 95*(2-3), 115–125.

Ranaivoson L., Naudin K., Ripoche A., Rabeharisoa L., Corbeels M. 2018. Is mulching an efficient way to control weeds? Effects of type and amount of crop residue in rainfed rice based cropping systems in Madagascar. *Field Crops Research, 217*: 20–31.

Roe, N.E., Stoffella, P.J. and Bryan, H.H., 1993. Municipal solid waste compost suppresses weeds in vegetable crop alleys. *HortScience, 28*(12), 1171–1172.

Saoir, S. and Mansfield, J., 1998. The potential for spent mushroom compost as a mulch for weed control in bramley orchards. In *International Conference on Integrated Fruit Production 525* (pp. 427–430).

Skroch, W.A., Powell, M.A., Bilderback, T.E. and Henry, P.H., 1992. Mulches: durability, aesthetic value, weed control, and temperature. *Journal of Environmental Horticulture, 10*(1), 43–45.

Splawski, C.E., Regnier, E.E., Harrison, S.K., Bennett, M.A. and Metzger, J.D., 2016. Weed suppression in pumpkin by mulches composed of organic municipal waste materials. *HortScience, 51*(6), 720–726.

Steinmetz Z., Wollmann C., Schaefer M., Buchmann C., David J., Tröger J., Muñoz K., Frör O., Schaumann, G.E., 2016. Plastic mulching in agriculture. Trading short-term agronomic benefits for long-term soil degradation? *Science of the Total Environment 550*, 690–705.

Swanton, C.J. and Weise, S.F., 1991. Integrated weed management: the rationale and approach. *Weed Technology, 5*(3), 657–663.

Tursun, D. Isik, Z. Demir, K. Jabran. 2018. Use of living, mowed, and soil-incorporated cover crops for weed control in apricot orchards. *Agronomy*, *8*, 150; https://doi.org/10.3390/agronomy8080150.

Yordanova, M. and Gerasimova, N., 2016. Effect of mulching on weed infestation and yield of beetroot (*Beta vulgaris* ssp. rapaceae atrorubra Krass). *Organic Agriculture*, *6*(2), 133–138.

Mulches for Insect Pest and Disease Management

Introduction

Insect pests and disease pathogens are included in the most important biotic stresses to agricultural crops. Although insecticides and fungicides are frequently applied for controlling insect pests and disease pathogens, there is need for alternative control methods in the wake of pesticide resistance evolution and environmental pollution (Barzman et al. 2015). Additionally, many of the insect pests or disease pathogens do not have a proper chemical control solution. Also, the alternative (non-chemical) control methods play an important role in integrated pest control (Barzman et al. 2015).

Mulches are famous for their benefits such as soil water retention, improving soil health, grain yield and quality, and weed suppression (Jabran et al. 2015a, b; Nawaz et al. 2017; Jabran and Chauhan 2018). Nevertheless, the mulches can also play a role in suppressing insect pests and disease pathogens (Brown and Tworkoski 2004; Farooq et al. 2011). Traditional use of mulches for suppressing diseases and insect pests in the conventional and organic farming is also on record (Poswal and Akpa 1991; Quintanilla-Tornel et al. 2016). Although residue from black plastic mulch may have some environmental implications, the straw mulch could be utilized in the insect pest management programs that are environmentally sustainable (Vincent et al. 2003).

Organic mulches mainly promote the activities, prevalence, and numbers of predators for an indirect control of insect pests. There is sufficient evidence that mulches contribute significantly to both the non-chemical and integrated pest management (Poswal and Akpa 1991; Brown and Tworkoski 2004). Hence, this chapter assesses the exploitation of various kind of mulches for suppressing the insect pests and reducing the disease incidences in various agricultural settings.

© The Author(s), under exclusive licence to Springer Nature Switzerland AG 2019
K. Jabran, *Role of Mulching in Pest Management and Agricultural Sustainability*, SpringerBriefs in Plant Science,
https://doi.org/10.1007/978-3-030-22301-4_3

Use of Mulches for Insect Pest Management

Organic mulches provide an environment that improves the insect pest control by facilitating the growth and prevalence of predators (Vincent et al. 2003; Flint 2018). The predators' density increases under the residue mulching that has been sustained for several seasons (Roger-Estrade et al. 2010; Quintanilla-Tornel et al. 2016). For instance, wheat straw mulch caused an indirect negative impact on the Colorado potato beetle (*Leptinotarsa decemlineata*) in the potato fields (Brust 1994). Within 15–20 days of this mulch application, there were significant populations of predators that could feed on the eggs as well as first and second instars of Colorado potato beetle. Potato defoliation was more than double in the fields without mulch and a yield decrease of more than 30%. A few important predators were *Chrysoperla carnea*, *Coleomegilla maculata*, and *Perillus bioculatus* (Brust 1994). Another impact of straw mulches may include provision of an alternative inhabitancy to the insect pests, and hence there is a lesser likeliness for pests to feed on crop while dwelling on the straw mulch. For example, the straw of sunn hemp (*Crotalaria juncea* L.) was applied as mulch in bush bean (*Phaseolus vulgaris* L.) to manage the disease lesser cornstalk borer (*Elasmopalpus lignosellus*) (Gill et al. 2010). The mulch decreased the disease incidence, increased the number of surviving plants, and enhanced the biomass and height of crop plants.

Compost mulch in the apple orchard not only supported the predators but also suppressed the insect pests including woolly apple aphid (*Eriosoma lanigerum*) and spotted tentiform leafminer (*Phyllonorycter blancardella*) (Brown and Tworkoski 2004). This insect suppression might had resulted from the physical characteristics, odor, or some chemical constituents of compost that were inappropriate for the well-being of insect pests.

Popularity of plastic mulches is increasing in agricultural production particularly in the controlled farming of high-value crops. Many of the plastic mulches can deflect or repel the insect pests (such as whiteflies, thrips, or aphids) through their color, odor, or surface characteristics (Vincent et al. 2003; Diaz and Fereres 2007). The variable light reflectance patterns of the differently colored plastic mulches could be exploited to deter the insect pests (Vincent et al. 2003). Zucchini squash (*Cucurbita pepo*) is an important horticultural commodity and has been infested by several insect pests and disease pathogens. Both the black plastic and silver spray could repel the aphid infestation in squash and increased the marketable squash yield over control (Summers et al. 1995) (Table 1).

Use of Mulches for Disease Management

Mulches positively modify the soil properties to prevent the prevalence of disease pathogens. The mulches also support the growth and activities of hyperparasites, microbial communities, and soil fauna that suppress the plant pathogens (Bonilla

Table 1 Effect of different types of mulches on control of insect pests or disease pathogens

Mulch type	Crop	Insect pest/ disease	Impact	References
Transparent plastic mulch	Cantaloupe (*Cucumis melo*)	Cucumber mosaic virus, aphids, and white fly	Decrease in disease incidence and populations of insect pests and increase in the fruit yield	Orozco et al. (1994)
Cotton gin trash	Tomato	Southern blight	Decrease in disease development	Liu et al. (2007)
Aluminum-coated plastic	Muskmelon (*Cucumis melo*)	Striped cucumber beetles	Decrease in pest density and increase in yield	Cline et al. (2008)
Straw mulch	Potato	Colorado potato beetle	A possible reduction in the insect pest infestation	Finckh et al. (2015)

et al. 2012). For instance, compost (or other manures) applied to rhizosphere positively change the activities and even composition of soil bacteria that can help to suppress the plant pathogens. Activities or growth of biocontrol agents can be enhanced with the application of organic mulches that will lead to natural suppression of many of the plant pests (Tiquia et al. 2002). Addition of mulch (or other organic amendments as well) may give the soil a disease-suppressive characteristic through improved soil properties and enhanced biological activities (Bonilla et al. 2012). Concentration and activities of enzymes (such as cellulase and laminarinase) are likely to be enhanced under the mulching that lead to reduced disease infection in the soil (Downer et al. 2001). For example, soil applied with *Sinapis alba* fresh residue, *Brassica carinata* pellets, sheep and horse manure (composted and semi-composted, respectively), and chicken litter (composted or semi-composted) had higher microbial respiration and concentrations of phosphatase, urease, β-glucosidase, and dehydrogenase (Núñez-Zofío et al. 2011). The soil was covered with a plastic mulch after these applications. *Phytophthora capsici*, an important disease of pepper, was decreased by more than 90% with *B. carinata* + *S. alba* treatment, and the other composted or semi-composted applications also caused a high decrease in the disease incidence (Núñez-Zofío et al. 2011). There was a negative correlation between dehydrogenase and *P. capsici* oospores. This is a soilborne disease, and chemical disinfection of soil has been an effective control method for such diseases; however, chemical disinfection is prohibited in the wake of its environmental damages. Hence, the non-chemical disease control methods such as application of mulches or organic amendments have become more important in this scenario (Olle et al. 2015).

In many parts of the world where avocado is grown, *Phytophthora cinnamomi*, a soilborne pathogen causes a root rot disease in the avocado orchards. A variety of control methods have been devised, yet the disease has devastating effects on avocado trees. Thick mulching of tree remains or other straw is helpful in suppressing *P. cinnamomi* in avocado through enhanced enzymatic (cellulase and laminarinase) activities (Downer et al. 2001). Wood mulches provide a cellulase enzyme activity of 25 U ml^{-1} or higher, and sporangia production in *P. cinnamomi* is reduced when exposed to 10–50 U ml^{-1} cellulase, while higher concentrations of the enzyme are needed to reduce vegetative biomass (Richter et al. 2011). White root rot is another important avocado disease (particularly in some parts of Asia and Europe) that is caused by *Rosellinia necatrix*. Almond shell and farm-yard waste are effective in suppressing this avocado disease through a different mechanism (Bonilla et al. 2015). The yard waste inhibits the disease through increased microbial populations and diversity and change in their composition, while almond shell impacts the disease by modifying the enzyme activities and bulk soil microbiota (Bonilla et al. 2015).

Different kinds of plastic mulches possess an interesting role in the control of various crop plant diseases. For example, thrips are vector for the tomato spotted wilt virus, and a UV-reflective mulch could reduce the incidence of this disease through reduced populations of thrips (Momol et al. 2002). Similarly, mulches such as silver spray and black plastic delayed the infection of viral diseases in squash (Summers et al. 1995). The disease incidence was greatly lowered through application of the mulches with almost a three-fourth increase in the squash yields (Summers et al. 1995). In another study, disease (apple scab and brown rot) incidence was reduced in the apple orchards after application of compost mulch (Brown and Tworkoski 2004).

Conclusions

Insect pests and disease suppression through mulches is particularly important for organic crop production (Flint 2018). Organic mulches may be used in agricultural production systems to support the predators and hence improve the insect pest control. This may have important implications for the environment-friendly pest control particularly in the agro-environments such as organic crop production. Repellent characteristics of colored plastic mulches can be exploited to avoid insect pests.

References

Barzman, M., Bàrberi, P., Birch, A.N.E., Boonekamp, P., Dachbrodt-Saaydeh, S., Graf, B., Hommel, B., Jensen, J.E., Kiss, J., Kudsk, P. and Lamichhane, J.R., 2015. Eight principles of integrated pest management. *Agronomy for Sustainable Development, 35*(4), 1199–1215.

Bonilla, N., Gutiérrez-Barranquero, J., Vicente, A. and Cazorla, F., 2012. Enhancing soil quality and plant health through suppressive organic amendments. *Diversity, 4*(4), 475–491.

Bonilla, N., Vida, C., Martínez-Alonso, M., Landa, B.B., Gaju, N., Cazorla, F.M. and de Vicente, A., 2015. Organic amendments to avocado crops induce suppressiveness and influence the composition and activity of soil microbial communities. *Applied and Environmental Microbiology, 81*(10), 3405–3418.

Brown, M.W. and Tworkoski, T., 2004. Pest management benefits of compost mulch in apple orchards. *Agriculture, Ecosystems & Environment, 103*(3), 465–472.

Brust, G.E., 1994. Natural enemies in straw-mulch reduce Colorado potato beetle populations and damage in potato. *Biological Control*, *4*(2), 163–169.

Cline, G.R., Sedlacek, J.D., Hillman, S.L., Parker, S.K. and Silvernail, A.F., 2008. Organic management of cucumber beetles in watermelon and muskmelon production. *HortTechnology*, *18*(3), 436–444.

Diaz, B.M. and Fereres, A., 2007. Ultraviolet-blocking materials as a physical barrier to control insect pests and plant pathogens in protected crops. *Pest Technology*, *1*(2), 85–95.

Downer, A.J., Menge, J.A. and Pond, E., 2001. Association of cellulytic enzyme activities in eucalyptus mulches with biological control of *Phytophthora cinnamomi*. *Phytopathology*, *91*(9), 847–855.

Farooq, M., Jabran, K., Cheema, Z.A., Wahid, A., Siddique, K.H.M., 2011. The role of allelopathy in agricultural pest management. *Pest Management Science*, *67*: 493–506

Finckh, M.R., Bruns, C., Bacanovic, J., Junge, S. and Schmidt, J.H., 2015. Organic potatoes, reduced tillage and mulch in temperate climates. *The Organic Grower*, *33*, 20–22.

Flint, M.L., 2018. *Pests of the garden and small farm: a grower's guide to using less pesticide* (Vol. 3332). UCANR Publications.

Gill, H.K., McSorley, R., Goyal, G. and Webb, S.E., 2010. Mulch as a potential management strategy for lesser cornstalk borer, *Elasmopalpus lignosellus* (Insecta: Lepidoptera: Pyralidae), in bush bean (*Phaseolus vulgaris*). *Florida Entomologist*, *93*(2), 183–191.

Jabran, K., Chauhan, B.S.. 2018. Non-Chemical Weed Control. Elsevier, Academic Press, London, United Kingdom.

Jabran, K., Ullah, E. and N. Akbar, 2015a. Mulching improves crop growth, grain length, head rice and milling recovery of basmati rice grown in water-saving production systems. *International Journal of Agriculture and Biology 17*: 920–928. https://doi.org/10.17957/IJAB/15.0019.

Jabran, K., Ullah, E., Hussain, M., Farooq, M., Yaseen, M., Zaman, U. and Chauhan, B.S., 2015b. Mulching improves water productivity, yield and quality of fine rice under water-saving rice production systems. *Journal of Agronomy and Crop Science*, *201*: 389–400. https://doi.org/10.1111/jac.12099.

Liu, B., Gumpertz, M.L., Hu, S. and Ristaino, J.B., 2007. Long-term effects of organic and synthetic soil fertility amendments on soil microbial communities and the development of southern blight. *Soil Biology and Biochemistry*, *39*(9), 2302–2316.

Momol, M.T., Funderburk, J.E., Olson, S. and Stavisky, J., 2002. Management of TSWV on tomatoes with UV-reflective mulch and acibenzolar-S-methyl. *Thrips and Tospoviruses*, 111–116. In R. Marullo and L. A. Mound [eds.], Proc. 7th Intl. Symp. on Thysanoptera, July 2-7, Reggio Calabria, Italy.

Nawaz, A., Farooq, M., Lal, R., Rehman, A., Hussain, T. and Nadeem, A., 2017. Influence of sesbania brown manuring and rice residue mulch on soil health, weeds and system productivity of conservation rice–wheat systems. *Land Degradation & Development*, *28*(3), 1078–1090.

Núñez-Zofío, M., Larregla, S. and Garbisu, C., 2011. Application of organic amendments followed by soil plastic mulching reduces the incidence of *Phytophthora capsici* in pepper crops under temperate climate. *Crop Protection*, *30*(12), 1563–1572.

Olle, M., Tsahkna, A., Tähtjärv, T. and Williams, I.H., 2015. Plant protection for organically grown potatoes–a review. *Biological Agriculture & Horticulture*, *31*(3), 147–157.

Orozco-S, M., Lopez-A, O., Perez-Z, O. and Delgadillo-S, F., 1994. Effect of transparent mulch, floating row covers and oil sprays on insect populations, virus diseases and yield of cantaloup. *Biological Agriculture & Horticulture*, *10*(4), 229–234.

Poswal, M.A.T. and Akpa, A.D., 1991. Current trends in the use of traditional and organic methods for the control of crop pests and diseases in Nigeria. *International Journal of Pest Management*, *37*(4), 329–333.

Quintanilla-Tornel, M.A., Wang, K.H., Tavares, J. and Hooks, C.R., 2016. Effects of mulching on above and below ground pests and beneficials in a green onion agroecosystem. *Agriculture, Ecosystems & Environment*, *224*, 75–85.

Richter, B.S., Ivors, K., Shi, W. and Benson, D.M., 2011. Cellulase activity as a mechanism for suppression of Phytophthora root rot in mulches. *Phytopathology*, *101*(2), 223–230.

Roger-Estrade, J., Anger, C., Bertrand, M. and Richard, G., 2010. Tillage and soil ecology: partners for sustainable agriculture. *Soil and Tillage Research, 111*(1), 33–40.

Summers, C.G., Stapleton, J.J., Newton, A.S., Duncan, R.A. and Hart, D., 1995. Comparison of sprayable and film mulches in delaying the onset of aphid-transmitted virus diseases in zucchini squash. *Plant Disease, 79*(11), 1126–1131.

Tiquia, S.M., Lloyd, J., Herms, D.A., Hoitink, H.A. and Michel Jr, F.C., 2002. Effects of mulching and fertilization on soil nutrients, microbial activity and rhizosphere bacterial community structure determined by analysis of TRFLPs of PCR-amplified 16S rRNA genes. *Applied Soil Ecology, 21*(1), 31–48.

Vincent, C., Hallman, G., Panneton, B. and Fleurat-Lessard, F., 2003. Management of agricultural insects with physical control methods. *Annual Review of Entomology, 48*(1), 261–281.

Mulches for Soil and Water Conservation

Introduction

Soil and water are among the factors that are inevitable for crop production. Both soil and water erosions cause great damage to the sustainability of arable fields. Erosion not only damages the soil structure but also makes the soil poor in nutrients (Issaka and Ashraf 2017). Runoff formed during the water erosion causes both the nutrient and soil losses (Oldeman et al. 2017; Smith et al. 2016). Several agronomic measures have been evaluated, and many of these were successful in restricting the water or soil erosion (Smith et al. 2016; Adimassu et al. 2017). Mulching of soil is one among such measures that could limit the soil or water erosion by decreasing the impact of wind on the soil, decreasing the influence of raindrops on the soil, and increasing the precipitation interception (Montenegro et al. 2013; Prosdocimi et al. 2016). Water conservation is particularly important for the arid areas because these have a low and irregular precipitation. Thus, a low water availability makes the farming in dry areas a difficult task. Recent events of climate change have made it inevitable to sustainability conserve soil and water.

If soil destruction is kept continued, the fertile lands being used for food production will soon become barren. There is need to exploit various kinds of soil covers (mulches) for soil protection and water conservation. Along with soil conservation, the mulches may positively impact the water retention in the soil, moisture evaporation from soil surface, soil moisture contents, and water infiltration in the soil. Hence, this chapter is aimed at explaining the role and mechanism of mulches in conserving the soil and water and the role of plastic, straw, and other types of mulches in conserving the water and soil resources.

K. Jabran, *Role of Mulching in Pest Management and Agricultural Sustainability*, SpringerBriefs in Plant Science, https://doi.org/10.1007/978-3-030-22301-4_4

Mechanism

Mulches play a role in conserving the soil and water resources through several of their positive impacts. The mulches block or at least reduce the intensity of moisture evaporation from the surface of soil. For example, maize straw mulching reduced the soil water evaporation by 21–40% compared to bare soil (Suying et al. 2005).

Mulches check the water and wind erosion by decreasing the impact of rain and wind on the soil and reducing the runoff (Prosdocimi et al. 2016). Another important impact of mulches includes the improved infiltration of water that reduces the soil and water loss through runoff. The infiltrated water is likely to be utilized by the plants over a long period of time. Further, the mulches also improve the soil water holding capacity through enhanced amount of organic matter. Mulches (particularly the organic mulches) also improve the soil properties and soil stability that indirectly helps to reduce the soil loss through wind or water erosion. For example, soil properties such as aggregate stability, total porosity, and bulk density are increased with an application of wheat straw mulch (Mulumba and Lal 2008). Straw mulch also increases the organic matter microbial biomass in the soil (García-Orenes et al. 2009).

Mulches for Water Conservation

Plastic Mulch for Water Conservation

Covering soil surface with mulch helps to conserve soil moisture contents. This technique can be employed to extend cropping to areas with limited water supply. In some parts of the world (particularly China and some African countries), plastic mulching has been combined with ridge-furrow to achieve conservation of water and soil. For instance, the practice (plastic mulch covering on a ridge-furrow) could increase soil water availability, root volume, water-use efficiency, plant dry matter production, and maize grain yield compared with a flat-sown crop without any soil covering (Mo et al. 2017). Ridge-furrow combined with a plastic covering has been evaluated for a water saving in the wheat crop in China, and this increased the water-use efficiency (~50%), yield components, and grain yield of wheat compared to control (flat planting) (Li et al. 2016).

Black plastic mulch (alone or combined with a cover crop) could increase the soil water status, reduce evapotranspiration and water-use efficiency, and increase apple yield in rainfed dryland areas (Zheng et al. 2017). Black or transparent plastic applied in potato fields could enhance the water-use efficiency and yield under drip irrigation system in China (Zhang et al. 2017). Among the three mulches (white and black plastic and rice straw), black plastic mulch was found to be the best in increasing leaf area index of tomato, fruit yield, and water productivity and decreasing evapotranspiration (Mukherjee et al. 2010). Among the degradable film and plastic mulches, the latter was more effective in decreasing evapotranspiration and increasing the yield and water-use efficiency of winter oilseed rape (*Brassica napus* L.) (Gu et al. 2017). Plastic mulch could decrease evaporation in the vineyard orchard (Pinamonti 1998).

Positive impacts of mulch application on water retention and grain yield have been noted in rice farming. An analysis of 36 rice farming sites indicated that covering the fields with mulch increased the rice grain yield by 18% (Liu et al. 2013). Mulching of rice soils with black plastic mulch could not only conserve the soil moisture but also increase the water productivity, grain quality, and paddy yield (Jabran et al. 2015a, b).

Straw Mulch for Water Conservation

Impact of straw mulch on soil moisture conservation or retention can be variable depending on the soil properties, type of crop, and environmental conditions. During summer season, the mulch of Egyptian clover was effective in maintaining soil water content and improving crop growth, but it did not perform better than the black plastic mulch (Jabran et al. 2015a, b). Contrarily, for wheat, a winter crop, rice husk provided a better soil water retention, water use, and yield gains than the plastic mulch (Chakraborty et al. 2008). The Egyptian clover mulch decreased the non-productive tillers and increased the water productivity in the water-saving rice systems (Jabran et al. 2015a).

Mulching with straw reduces the water evaporation from soil. For instance, enhanced water storage and a reduced evaporation were resulted with application of wheat straw as a mulch (2–4 t/ha) (Ji and Unger 2001). Different kinds of straw mulches (grass clippings, wheat, and leaf debris) were equally effective in reducing the water evaporation from soil surface, and a 5 cm depth of these mulches could decrease the soil surface moisture evaporation by about 40% (McMillen 2013). An increase in mulch depth to 10 cm enhanced the soil moisture by 10%, but a further increase in the mulch depth (up to 15 cm) did not bring a further advantage (McMillen 2013). Wheat straw application (2–16 t/ha) could enhance both the moisture retention in the soil and its availability (Mulumba and Lal 2008).

Wheat straw and black plastic mulches could ameliorate the impacts of drought stress on the cucumber plants through decreased evaporation (Kirnak and Demirtas 2006). The mulches not only improved the water-use efficiency of cucumber but also increased its leaf area and biomass production. Further, a positive impact of mulches on the size of fruit and plant nutrient uptake was also observed (Kirnak and Demirtas 2006).

Other Mulches for Water Conservation

Although plastic and straw may be considered as the mulches that are mostly evaluated in scientific studies or applied in the agricultural settings for water conservation, several other types of mulches also have a potential to be utilized for water conservation. Compost may be an example of such types of mulches that have the potential not only to improve the soil moisture holding capacity but also improve other physical and chemical properties of soil. Compost mulch decreases evaporation, increases water permeability, and enhances soil water storage in vineyard orchards (Pinamonti 1998). Mulch of compost made from municipal waste increases the percolation of rainwater by 85% and effectively reduces the runoff (Agassi et al. 1998).

Sand gravel and volcanic residue constitute unique types of mulches that have been used in soil and water conservation. A traditional way of conserving water and soil in the northwest China includes use of sand and gravel as a mulch, and the method has been in use for centuries (Li 2003). Tephra (volcanic residue) was highly effective in soil moisture conservation and could increase the water infiltration and reduce evaporation (Tejedor et al. 2003). This mulch enhanced the water retention to double at a 1 m depth and to eight times in the surface layer (Tejedor et al. 2003) (Table 1).

Table 1 Impact of different types of mulches on the soil water content

Mulch type	Impact on soil water	References
Plastic mulch	Increased soil water contents	Li et al. (2012)
	Increase in soil moisture contents and water productivity	Jabran et al. (2015a, b)
	Improved maize yields through enhanced moisture availability and modified soil temperature conditions	Lee et al. (2019)
Degradable film mulch	Increased soil water contents	Li et al. (2012)
Straw mulch	Increased soil water contents	Li et al. (2012)
	Enhanced soil water storage	Jun et al. (2014)
	Increase in soil moisture contents	Jabran et al. (2015a, b)
	Reduced water needs and enhanced water productivity	Jat et al. (2015)
Gravel mulch	Increase in soil water content	Jun et al. (2014)
Compost made from municipal waste	Increase (85%) in water percolation	Agassi et al. (1998)
Oat straw and olive twigs as mulches	Decrease in loss of rainfall water	García-Orenes et al. (2009)
Transparent plastic mulch	Reduced evapotranspiration	Wang et al. (2009)

Mulches for Soil Conservation

Stubble and Straw Mulches for Soil Conservation

Straw mulch prevents the soil losses by providing a physical covering to the soil and also improving the soil properties such as improved soil aggregate stability and enhanced organic matter (García-Orenes et al. 2009).

Wildfires are likely to cause deforestation, and a post-wildfire degradation of land resources is common. This land degradation includes damaged soil physical, microbial, and chemical properties. Mulching with straw can help to prevent the post-wildfire land degradation or to rehabilitate the soil that have already faced a damage. The positive impacts (including a reduction in soil erosion) of mulch on soil stabilization are higher than seeding a crop on a degraded soil (Díaz-Raviña et al. 2012). Nevertheless, mulching for short term can be useful in preventing the immediate soil losses, but for the improved soil quality or properties, mulching for longer durations is required. Straw mulch (200 g/m^2) was applied to restore and conserve the soil after a fire regime (Bautista et al. 1996). Mulching conserved the soil by declining sediment loss (>7 times), runoff, and crusting and also caused an increase in vegetation growth (Bautista et al. 1996).

Conclusions

The discussion in this chapter concludes that different kinds of mulches provide a great opportunity to conserve soil and water under a variety of situations. This is important in the perspective of sustainable use of these resources. Various kinds of mulches reduce the evaporation of moisture from surface of soil, increase the water infiltration into soil, increase water productivity and water-use efficiency, enhance soil water holding capacity, and stabilize the soil by enhancing organic matter in it. Although, mulches possess a potential to conserve soil and water, their high costs are likely to limit their use. Nowadays, this is a great challenge to manage or recycle the wastes originating from difference sources. Out of these wastes, the organic (or even the non-organic) ones that contain no heavy metal or other harmful material may be modified into mulches, and these mulches can then be employed to cover the soil in order to avoid or decrease the wind and water erosion of soil. Hence, converting waste materials into mulches can provide low-cost raw material for preparation of mulches.

References

Adimassu, Z., Langan, S., Johnston, R., Mekuria, W. and Amede, T., 2017. Impacts of soil and water conservation practices on crop yield, run-off, soil loss and nutrient loss in Ethiopia: review and synthesis. *Environmental Management*, 59(1), 87–101.

Agassi, M., Hadas, A., Benyamini, Y., Levy, G.J., Kautsky, L., Avrahamov, L. and Zhevelev, H., 1998. Mulching effects of composted MSW on water percolation and compost degradation rate. *Compost Science & Utilization*, 6(3), 34–41.

Bautista, S., Bellot, J. and Vallejo, V.R., 1996. Mulching treatment for postfire soil conservation in a semiarid ecosystem. *Arid Land Research and Management*, 10(3), 235–242.

Chakraborty, D., Nagarajan, S., Aggarwal, P., Gupta, V.K., Tomar, R.K., Garg, R.N., Sahoo, R.N., Sarkar, A., Chopra, U.K., Sarma, K.S. and Kalra, N., 2008. Effect of mulching on soil and plant water status, and the growth and yield of wheat (*Triticum aestivum* L.) in a semi-arid environment. *Agricultural Water Management*, 95(12), 1323–1334.

Díaz-Raviña, M., Martín, A., Barreiro, A., Lombao, A., Iglesias, L., Díaz-Fierros, F. and Carballas, T., 2012. Mulching and seeding treatments for post-fire soil stabilisation in NW Spain: short-term effects and effectiveness. *Geoderma*, 191, 31–39.

García-Orenes, F., Cerdà, A., Mataix-Solera, J., Guerrero, C., Bodí, M.B., Arcenegui, V., Zornoza, R. and Sempere, J.G., 2009. Effects of agricultural management on surface soil properties and soil–water losses in eastern Spain. *Soil and Tillage Research*, 106(1), 117–123.

Gu, X.B., Li, Y.N. and Du, Y.D., 2017. Biodegradable film mulching improves soil temperature, moisture and seed yield of winter oilseed rape (*Brassica napus* L.). *Soil and Tillage Research*, 171, 42–50.

Issaka, S. and Ashraf, M.A., 2017. Impact of soil erosion and degradation on water quality: a review. *Geology, Ecology, and Landscapes*, 1(1), 1–11.

Jabran, K., Ehsan Ullah and N. Akbar, 2015b. Mulching improves crop growth, grain length, head rice and milling recovery of basmati rice grown in water-saving production systems. *International Journal of Agriculture and Biology*, 17, 920–928. https://doi.org/10.17957/IJAB/15.0019.

Jabran, K., Ehsanullah, M. Hussain, M. Farooq, M. Yaseen, U. Zaman and B.S. Chauhan, 2015a. Mulching improves water productivity, yield and quality of fine rice under water-saving rice production systems. *Journal of Agronomy and Crop Science*, 201, 389–400. https://doi.org/10.1111/jac.12099.

Jat, H.S., Singh, G., Singh, R., Choudhary, M., Jat, M.L., Gathala, M.K. and Sharma, D.K., 2015. Management influence on maize–wheat system performance, water productivity and soil biology. *Soil Use and Management*, 31(4), 534–543.

Ji, S. and Unger, P.W., 2001. Soil water accumulation under different precipitation, potential evaporation, and straw mulch conditions. *Soil Science Society of America Journal*, 65(2), 442–448.

Jun, F., Yu, G., Quanjiu, W., Malhi, S.S. and Yangyang, L., 2014. Mulching effects on water storage in soil and its depletion by alfalfa in the Loess Plateau of northwestern China. *Agricultural Water Management*, 138, 10–16.

Kirnak, H. and Demirtas, M.N., 2006. Effects of different irrigation regimes and mulches on yield and macronutrition levels of drip-irrigated cucumber under open field conditions. *Journal of Plant Nutrition*, 29(9), 1675–1690.

Lee, J.G., Hwang, H.Y., Park, M.H., Lee, C.H. and Kim, P.J., 2019. Depletion of soil organic carbon stocks are larger under plastic film mulching for maize. *European Journal of Soil Science*. https://doi.org/10.1111/ejss.12757.

Li, C., Wen, X., Wan, X., Liu, Y., Han, J., Liao, Y. and Wu, W., 2016. Towards the highly effective use of precipitation by ridge-furrow with plastic film mulching instead of relying on irrigation resources in a dry semi-humid area. *Field Crops Research*, 188, 62–73.

Li, R., Hou, X., Jia, Z., Han, Q. and Yang, B., 2012. Effects of rainfall harvesting and mulching technologies on soil water, temperature, and maize yield in Loess Plateau region of China. *Soil Research*, 50(2), 105–113.

Li, X.Y., 2003. Gravel–sand mulch for soil and water conservation in the semiarid loess region of northwest China. *Catena, 52*(2), 105–127.

Liu, M., Lin, S., Dannenmann, M., Tao, Y., Saiz, G., Zuo, Q., Sippel, S., Wei, J., Cao, J., Cai, X. and Butterbach-Bahl, K., 2013. Do water-saving ground cover rice production systems increase grain yields at regional scales?. *Field Crops Research, 150*, 19–28.

McMillen, M., 2013. The effect of mulch type and thickness on the soil surface evaporation rate. Horticulture and Crop Science Department. California Polytechnic State University, San Luis Obispo USA.

Mo, F., Wang, J.Y., Zhou, H., Luo, C.L., Zhang, X.F., Li, X.Y., Li, F.M., Xiong, L.B., Kavagi, L., Nguluu, S.N. and Xiong, Y.C., 2017. Ridge-furrow plastic-mulching with balanced fertilization in rainfed maize (*Zea mays* L.): An adaptive management in east African Plateau. *Agricultural and Forest Meteorology, 236*, 100–112.

Montenegro, A.A.A., Abrantes, J.R.C.B., de Lima, J.L.M.P., Singh, V.P., Santos, T.E.M., 2013. Impact of mulching on soil and water dynamics under intermittent simulated rainfall. *Catena, 109*, 139–149

Mukherjee, A., Kundu, M. and Sarkar, S., 2010. Role of irrigation and mulch on yield, evapotranspiration rate and water use pattern of tomato (*Lycopersicon esculentum* L.). *Agricultural Water Management, 98*(1), 182–189.

Mulumba, L.N. and Lal, R., 2008. Mulching effects on selected soil physical properties. *Soil and Tillage Research, 98*(1), 106–111.

Oldeman, L.R., Hakkeling, R.T.A. and Sombroek, W.G., 2017. *World map of the status of human-induced soil degradation: an explanatory note.* International Soil Reference and Information Centre. Nairobi: United Nations Environment Programme (UNEP).

Pinamonti, F., 1998. Compost mulch effects on soil fertility, nutritional status and performance of grapevine. *Nutrient Cycling in Agroecosystems, 51*(3), 239–248.

Prosdocimi, M., Jordán, A., Tarolli, P., Keesstra, S., Novara, A. and Cerdà, A., 2016. The immediate effectiveness of barley straw mulch in reducing soil erodibility and surface runoff generation in Mediterranean vineyards. *Science of the Total Environment, 547*, 323–330.

Smith, P., House, J.I., Bustamante, M., Sobocká, J., Harper, R., Pan, G., West, P.C., Clark, J.M., Adhya, T., Rumpel, C. and Paustian, K., 2016. Global change pressures on soils from land use and management. *Global Change Biology, 22*(3), 1008–1028.

Suying, C., Xiying, Z., Dong, P. and Hongyong, S., 2005. Effects of corn straw mulching on soil temperature and soil evaporation of winter wheat field. *Transactions of The Chinese Society of Agricultural Engineering, 10*, 038.

Tejedor, M., Jiménez, C. and Dıaz, F., 2003. Volcanic materials as mulches for water conservation. *Geoderma, 117*(3–4), 283–295.

Wang, F.X., Feng, S.Y., Hou, X.Y., Kang, S.Z. and Han, J.J., 2009. Potato growth with and without plastic mulch in two typical regions of Northern China. *Field Crops Research, 110*(2), 123–129.

Zhang, Y.L., Wang, F.X., Shock, C.C., Yang, K.J., Kang, S.Z., Qin, J.T. and Li, S.E., 2017. Influence of different plastic film mulches and wetted soil percentages on potato grown under drip irrigation. *Agricultural Water Management, 180*, 160–171.

Zheng W, Wen M, Zhao Z, Liu J, Wang Z, Zhai B, et al., (2017) Black plastic mulch combined with summer cover crop increases the yield and water use efficiency of apple tree on the rainfed Loess Plateau. *PLoS ONE, 12*(9), e0185705. https://doi.org/10.1371/journal.pone.0185705.

Mulches for Enhancing Biological Activities in Soil

Introduction

Biological activities in the soil are inevitable to maintain soil health, nutrient transformations (to a form usable by plants), and hence sustain the plant growth. Living component of soil consists of flora including algae, fungi and bacteria, and fauna including nematodes, protozoa, earthworms, termites, microarthropods, arthropods, enchytraeid, etc. (Lal 1988; Roger-Estrade et al. 2010). Fungi, bacteria, and other living components of soil carry out organic matter decomposition, nutrient transformation, nutrient storage, and aggregation of soil particles and play a role in nitrogen and carbon cycles and several other processes (Waid 1999; Potthoff et al. 2005). The rhizosphere part of the soil possesses the highest biological activities particularly those of microorganisms and soil macrofauna. Importance and details of metabiosis have been described by Waid (1999).

   The soils that witness a biological activity do possess a higher concentration of nutrients and organic matter and improved texture, porosity, and infiltration (Lal 1988; Paz-Ferreiro and Fu 2016). Soil biological activities are mostly affected by factors such as soil type, soil composition, plant type, tillage, fertilization, crop rotation, and soil amendments (particularly the organic amendments) (Bonilla et al. 2012). The role of mulches is well-known in improving the biological activities in soil (Lal 1988). Mulches that comprise of organic materials increase the microbial and subsequently the enzymatic (particularly dehydrogenase as it indicates organic materials' microbial oxidation) activity in soil. The enhanced microbial activities are usually measured as an increase in soil respiration or enzymatic (β-glucosidase, dehydrogenase, urease, phosphatase, etc.) activities. Addition of organic matter in the form of mulches (along with other benefits) facilitates and promotes the soil flora and fauna (Lal 1988). Due to role of mulches in improving soil health, conservation agriculture also stresses the need for leaving crop residue to soil for improving its physical properties as well as the enhanced biological activities (Farooq

K. Jabran, *Role of Mulching in Pest Management and Agricultural Sustainability*, SpringerBriefs in Plant Science, https://doi.org/10.1007/978-3-030-22301-4_5

et al. 2011; Kahlon et al. 2013). Nevertheless, the high-quality mulch (e.g., the ones with a narrower C/N ratio) is better to support the population, growth, and activity of soil living organisms. For example, density of earthworms was impacted not only by the presence of mulch but also by the quality of mulch, and the high-quality mulch possessed the higher number of earthworms than the low-quality mulch (Tian et al. 1997). Although mulches are considered to improve biological activities in a soil, their impact is mostly high only in the topsoil particularly in the initial times of mulch application (Yang et al. 2003).

For improving soil productivity and increasing sustainability of food production, there should be cultural practices that facilitate the beneficial soil organisms and microorganisms and suppress the plant pests as well (Abawi and Widmer 2000). Hence, organic mulches are desired to be added to the soil for enhanced biological activities, i.e., those of beneficial soil flora and fauna. This chapter is aimed to discuss the role of mulches in supporting the soil biological activities and subsequent improvement in enzymatic activities, the soil properties, and the crop productivity.

Effect of Straw Mulch on Biological Activities in Soil

Mulching of plant residue for long times enhances diversity of spiders and ground beetles (Roger-Estrade et al. 2010). *Crotalaria retusa* L. and *Brachiaria ruziziensis* Germain and Ewar mulching in no-till cotton facilitates the rise in populations and activities of staphylinids, termites, spiders, ants, carabids, and centipedes (Brévault et al. 2007). Mulching combined with no-till enhances the biomass of soil macrofauna (Blanchart et al. 2007). Some mulches (*Pennisetum pedicellatum* straw and woody parts of *Pterocarpus lucens*) were evaluated for rehabilitation of a crusted soil (Mando and Stroosnijder 1999). Termites were the major biological agents that had a higher density in the mulched plots and helped to rehabilitate the soil (Mando and Stroosnijder 1999). Similarly, eucalyptus mulching improved the fungal populations and microbial and enzymatic activities that could suppress the *Phytophthora* disease (Downer et al. 2001). Yard waste from eucalyptus increased the diversity of bacteria, but the microbial activity was mainly restricted in the topsoil (Yang et al. 2003). Populations of azotobacter, fungi, actinomycetes, and bacteria were enhanced in green gram and maize with the application of wheat straw mulch (Gaur and Mukherjee 1980).

Rise in soil respiration and enzyme activities is an indication of the enhanced microbial activities. Such enzymatic activities are important for catalyzing various reactions in the soil including those of nutrient transformations and decomposition of organic matter, etc. (Burns et al. 2013). Macromolecules such as cellulose, lignin, hemicellulose, chitin, pectin, etc. do require a microbial enzymatic activity for their recycling. Many studies demonstrate that adding straw mulch (or other organic amendments) improves the soil respiration or enzymatic activities (Zhang et al. 2015). The increased enzymatic activities in turn improve the soil quality and several of other soil properties and impact the crop growth positively.

A 9–12 t/ha of wheat straw mulch was applied in the maize field and this treatment increased the quantities of microbial organisms in the field including those of *Azotobacter*, fungi, and actinomycetes; the enzymatic activities (invertase, urease, dehydrogenase, protease) were also increased (Zhang et al. 2015). A ~9% increase in maize yield along with increased photosynthesis and stomatal conductance were also observed. In another study, mulch of maize stover combined with no-till helped to increase the abundance of soil macrofauna and supported to establish predators and decomposers (Jiang et al. 2018) (Table 1).

Effect of Compost and Other Mulch on Biological Activities in Soil

Composts are rich in nutrients; these not only enhance the soil fertility but also help increasing the soil microbial activities. For instance, higher microbial respiration, populations, and biocontrol organisms were found in the soil that was applied with a compost mulch compared with the non-mulched soil (Tiquia et al. 2002).

Soil covering with materials other than straw or compost may also play a role in enhancing biological activities in soil. Plastic mulch applied in a continuous

Table 1 Effect of various mulches on the biological activities in soils of different crops

Mulch type	Crop	Biological activities	References
Gliricidia sepium, Leucaena leucocephala	–	More than 50% increase in earthworm populations	Tian et al. (1997)
Leaves maple (*Acer* spp.)	Perennial ryegrass (*Lolium perenne*)	Increased dehydrogenase and β-glucosidase	Acosta-Martinez et al. (1999)
Shredded paper and municipal compost	Apple orchard	Increase in number of bacterivorous nematodes and protozoa	Forge et al. (2003)
Plastic mulch	Spring wheat	Increase in soil microbial biomass	Li et al. (2004)
Wheat straw mulch	Organic tomato	Increase in soil microbial biomass and respiration	Tu et al. (2006)
Brassicaceae seed meal	Organic apple orchard	Increase in nematode density	Hoagland et al. (2008)
Wheat straw mulch	Soybean	Increased nitrogenase activity	Siczek and Lipiec (2011)
Maize stalk mulch	Apple	Change in bacteria communities	Chen et al. (2014)
Ryegrass (*Lolium multiflorum* L.)	Rice	Increase in alkaline phosphatase, β-glucosidase, arylsulfatase, and arylamidase activities	Hai-Ming et al. (2014)
Almond shell mulch	Avocado	Increase in activity of dehydrogenase, phosphomonoesterase, and protease	López et al. (2014)

maize cropping system enhanced biological activities in the soil; this included ~20–24% increase in the number of bacteria and actinomyces and ~12–14% increase in fungi count, along with enhanced urease activity and soil respiration (Xishi et al. 1998). Many of the materials that are available easily and without extra costs may be evaluated for their positive impacts on the biological activities in soil. However, it should be ensured that these mulch materials do not contain heavy materials or other pollutants.

Conclusions

Mulches (particularly the organic mulches) possess a strong positive impact on the biological activities in soil. The recent literature provides sufficient evidence that mulches have a role in increasing the earthworm populations, enzymatic activities in the soil, and the microbial biomass in the soil. Importantly, the concentrations and activities of enzymes such as β-glucosidase, dehydrogenase, cellobiohydrolase, urease, phosphatase, xylanase, phosphomonoesterase, protease, etc. are enhanced under organic mulches, and this has positive implications for several of the processes in the soil. Further, the mulches also facilitate the prevalence of hyperparasites, which are important for natural pest control.

References

Abawi, G.S. and Widmer, T.L., 2000. Impact of soil health management practices on soilborne pathogens, nematodes and root diseases of vegetable crops. *Applied Soil Ecology*, 15(1), 37–47.

Acosta-Martinez, V., Reicher, Z., Bischoff, M. and Turco, R.F., 1999. The role of tree leaf mulch and nitrogen fertilizer on turfgrass soil quality. *Biology and Fertility of Soils*, 29(1), 55–61.

Blanchart, E., Bernoux, M., Sarda, X., Siqueira Neto, M., Cerri, C.C., Piccolo, M., Douzet, J.M., Scopel, E. and Feller, C., 2007. Effect of direct seeding mulch-based systems on soil carbon storage and macrofauna in Central Brazil. *Agriculturae Conspectus Scientificus*, 72(1), 81–87.

Bonilla, N., Gutiérrez-Barranquero, J., Vicente, A. and Cazorla, F., 2012. Enhancing soil quality and plant health through suppressive organic amendments. *Diversity*, 4(4), 475–491.

Brévault, T., Bikay, S., Maldès, J.M. and Naudin, K., 2007. Impact of a no-till with mulch soil management strategy on soil macrofauna communities in a cotton cropping system. *Soil and Tillage Research*, 97(2), 140–149.

Burns, R.G., DeForest, J.L., Marxsen, J., Sinsabaugh, R.L., Stromberger, M.E., Wallenstein, M.D., Weintraub, M.N. and Zoppini, A., 2013. Soil enzymes in a changing environment: current knowledge and future directions. *Soil Biology and Biochemistry*, 58, 216–234.

Chen, Y., Wen, X., Sun, Y., Zhang, J., Wu, W. and Liao, Y., 2014. Mulching practices altered soil bacterial community structure and improved orchard productivity and apple quality after five growing seasons. *Scientia Horticulturae*, 172, 248–257.

Downer, A.J., Menge, J.A. and Pond, E., 2001. Association of cellulytic enzyme activities in eucalyptus mulches with biological control of *Phytophthora cinnamomi*. *Phytopathology*, 91(9), 847–855.

Farooq, M., K.C. Flower, K. Jabran, A. Wahid, K.H.M. Siddique. 2011. Crop yield and weed management in rainfed conservation agriculture. *Soil and Tillage Research,* 117, 172–183.

Forge, T.A., Hogue, E., Neilsen, G. and Neilsen, D., 2003. Effects of organic mulches on soil microfauna in the root zone of apple: implications for nutrient fluxes and functional diversity of the soil food web. *Applied Soil Ecology*, 22(1), 39–54.

Gaur, A.C. and Mukherjee, D., 1980. Recycling of organic matter through mulch in relation to chemical and microbiological properties of soil and crop yields. *Plant and Soil*, 56(2), 273–281.

Hai-Ming, T., Xiao-Ping, X., Wen-Guang, T., Ye-Chun, L., Ke, W. and Guang-Li, Y., 2014. Effects of winter cover crops residue returning on soil enzyme activities and soil microbial community in double-cropping rice fields. *PLoS One*, 9(6), e100443.

Hoagland, L., Carpenter-Boggs, L., Granatstein, D., Mazzola, M., Smith, J., Peryea, F. and Reganold, J.P., 2008. Orchard floor management effects on nitrogen fertility and soil biological activity in a newly established organic apple orchard. *Biology and Fertility of Soils*, 45(1), 11.

Jiang, Y., Ma, N., Chen, Z. and Xie, H., 2018. Soil macrofauna assemblage composition and functional groups in no-tillage with corn stover mulch agroecosystems in a mollisol area of northeastern China. *Applied Soil Ecology*, 128, 61–70.

Kahlon, M.S., Lal, R. and Ann-Varughese, M., 2013. Twenty two years of tillage and mulching impacts on soil physical characteristics and carbon sequestration in Central Ohio. *Soil and Tillage Research*, 126, 151–158.

Lal, R., 1988. Effects of macrofauna on soil properties in tropical ecosystems. *Agriculture, Ecosystems & Environment*, 24(1–3), 101–116.

Li, F.M., Song, Q.H., Jjemba, P.K. and Shi, Y.C., 2004. Dynamics of soil microbial biomass C and soil fertility in cropland mulched with plastic film in a semiarid agro-ecosystem. *Soil Biology and Biochemistry*, 36(11), 1893–1902.

López, R., Burgos, P., Hermoso, J.M., Hormaza, J.I. and González-Fernández, J.J., 2014. Long term changes in soil properties and enzyme activities after almond shell mulching in avocado organic production. Soil and Tillage Research, 143, 155–163.

Mando, A. and Stroosnijder, L., 1999. The biological and physical role of mulch in the rehabilitation of crusted soil in the Sahel. *Soil Use and Management*, 15(2), 123–127.

Paz-Ferreiro, J. and Fu, S., 2016. Biological indices for soil quality evaluation: perspectives and limitations. *Land Degradation & Development*, 27(1), 14–25.

Potthoff, M., Dyckmans, J., Flessa, H., Muhs, A., Beese, F. and Joergensen, R.G., 2005. Dynamics of maize (*Zea mays* L.) leaf straw mineralization as affected by the presence of soil and the availability of nitrogen. *Soil Biology and Biochemistry*, 37(7), 1259–1266.

Roger-Estrade, J., Anger, C., Bertrand, M. and Richard, G., 2010. Tillage and soil ecology: partners for sustainable agriculture. *Soil and Tillage Research*, 111(1), 33–40.

Siczek, A. and Lipiec, J., 2011. Soybean nodulation and nitrogen fixation in response to soil compaction and surface straw mulching. *Soil and Tillage Research*, 114(1), 50–56.

Tian, G., Kang, B.T. and Brussaard, L., 1997. Effect of mulch quality on earthworm activity and nutrient supply in the humid tropics. *Soil Biology and Biochemistry*, 29(3–4), 369–373.

Tiquia, S.M., Lloyd, J., Herms, D.A., Hoitink, H.A. and Michel Jr, F.C., 2002. Effects of mulching and fertilization on soil nutrients, microbial activity and rhizosphere bacterial community structure determined by analysis of TRFLPs of PCR-amplified 16S rRNA genes. *Applied Soil Ecology*, 21(1), 31–48.

Tu, C., Ristaino, J.B., Hu, S. 2006. Soil microbial biomass and activity in organic tomato farming systems: Effects of organic inputs and straw mulching. *Soil Biology and Biochemistry*, 38, 247–255.

Waid, J.S., 1999. Does soil biodiversity depend upon metabiotic activity and influences?. *Applied Soil Ecology*, 13(2), 151–158.

Xishi, C., Shufan, G., Jingkuan, W. and Jian, Z., 1998. Effect of mulching cultivation with plastic film on soil microbial population and biological activity. *Chinese Journal of Applied Ecology*, 4, 435–439.

Yang, Y.J., Dungan, R.S., Ibekwe, A.M., Valenzuela-Solano, C., Crohn, D.M. and Crowley, D.E., 2003. Effect of organic mulches on soil bacterial communities one year after application. *Biology and Fertility of Soils*, 38(5), 273–281.

Zhang, X., Qian, Y. and Cao, C., 2015. Effects of straw mulching on maize photosynthetic characteristics and rhizosphere soil micro-ecological environment. *Chilean Journal of Agricultural Research*, 75(4), 481–487.

Mulches for Regulation of Soil Temperature

Introduction

Temperature is among the most important factors that impact all the growth stages of crop plants. Seeds of different plant species require different ranges of temperature to start and accomplish the process of their germination (Rosbakh and Poschlod 2015). Synthesis of food in plants (photosynthesis) is directly impacted by the temperature conditions in the surrounding environment (Kumarathunge et al. 2019). Temperature extremes can cause several of negative impacts on the growth of plants such as production of reactive oxygen species (Zhu 2016; Szymańska et al. 2017). Recently, the climate change including global warming is worsening the negative impacts of temperature extremes on plants and different plant ecosystems (Nolan et al. 2018).

A wide range of agronomic, ecological, chemical, and biotechnological approaches have been used to tackle the negative effects of temperature stress on plants. One among the options is to use soil covering materials that eventually modify soil temperature and its microclimate positively. Previous literature clearly indicates the benefits of using mulching materials for positively modifying the soil temperature. For example, a comparison between red and black plastic mulches showed that black plastic mulch provided a slightly higher soil temperature (only 0.2 °C) than the red one (Kasperbauer 2000). Among the black, white, and black-white mulches too, the black plastic mulches caused greater increase in root temperature that had a negative impact on tuber yield and nitrogen metabolism (Ruiz et al. 1999). As fluctuations in daily temperature could have severe negative impacts not only on the plants but also on the other biological communities of the soil, hence, this chapter is aimed at explaining the role of different types of mulches in regulating the soil temperature. Further the mechanism of soil temperature regulation by mulches and benefits of temperature regulation by the mulches to the plants have also been discussed in this chapter.

Mechanism and Importance

Mulches raise the soil temperature owing to absence of an exchange of latent heat, decrease in outgoing long-wave radiations, and the weakened exchange of sensible heat (Genyun 1991). This is particularly true for the plastic mulch and partly for the straw mulch. It is well-known that straw mulches are in form of pieces which does not restrict air and heat circulation, whereas the plastic mulches are in form of a single or a few pieces for whole of the field. Usually, this has been noted that straw mulches lower the soil temperature, while a vice versa impact has been observed in case of plastic mulches. The straw mulches reduce the direct impact of sunlight on the soil and do not tightly trap the heat, whereas the colored (particularly black) plastic mulches not only trap the heat but also reduce the circulation of air and exchange of heat with the air. This way, the straw and plastic mulches have differential impact on the soil temperature regulation.

Plastic Mulch for Regulating Soil Temperature

Crops like cotton require a warm soil during the germination, growth of the seedlings, and their establishment. A poor germination for cotton is frequently witnessed under the cold soil conditions. Chilling stress during germination and seedling stage has also been observed for the maize crop (Farooq et al. 2008, 2017). Further, the swung in temperature during the germination and seedling growth stages could also weaken the establishment of crop stand. In this wake, plastic mulch may act as an excellent insulation to limit temperature fluctuations as well as raise the soil temperature reasonably to improve the germination of seeds and establishment of crop stand (Wang et al. 2011).

Black plastic mulch consistently raised the soil temperature in maize crop sown under the ridge-furrow system, and maize grain yield was 13% higher with this mulch compared to control (bare soil) (Li et al. 2013). In tomato crop too, both the black and white plastic mulches raised the soil temperature greater than control and other mulches (Decoteau et al. 1989; Fortnum et al. 1997). In the pickling cucumber cultivation, the soil temperatures including the average, maximum, and minimum were raised by silver/black and black mulches, and this supported the plants in getting higher yields compared to mulch-less soil (Torres-Olivar et al. 2018).

Elevated soil temperature through plastic mulching may enhance the crop emergence. For instance, the use of transparent plastic mulch in potato fields raised the topsoil temperature by ~3 °C during the crop emergence period (Zhao et al. 2012). This rise in soil temperature accelerated the crop emergence and extended the crop duration that resulted in increased plant biomass, leaf area and height, and increased tuber yield. The mulching also bettered the soil moisture status and enhanced the water use efficiency of potato. There was a slight decrease in soil temperature with the mulch application toward the end of the growing season that could be ameliorated by removing the mulch from the field during that period (Zhao et al. 2012). Transparent plastic mulch applied in maize crop could reduce the diurnal temperature fluctuations and increase the soil surface temperature along with improved maize yields, economic returns, and water use efficiency (Zhang et al. 2019).

Although several of the studies demonstrated that black or even other colored plastic mulches could regulate the soil temperature, contrary results have also been observed. For instance, black plastic mulch was ineffective against soil temperature fluctuations in grapevine (Pinamonti 1998). However, evidence for the ineffectiveness of plastic mulches in soil temperature regulation is rare, and most of the studies indicate the importance of mulching in soil temperature regulation (Table 1). An important issue is that overheating caused by the plastic mulch can damage the plant tissues (particularly plant roots) as well as other biological entities in the soil. For example, the rise in soil temperature by the transparent plastic mulch could decrease the crop emergence and damage the potato crop during the hot weather conditions (Wang et al. 2009). This implies that plastic (transparent, black, or other colored) mulch should be used in areas where (along with other factors such as need for weed control, soil and water conservations, or others) heating of soil is also required in addition to temperature regulation.

Straw Mulch for Regulating Soil Temperature

Mulching the soil with straw reduces the soil temperature during summer season as well as narrows the temperature amplitudes during a day. For example, maize straw mulch was tested for temperature regulation in the wheat crop (Suying et al. 2005). The mulch not only decreased the soil temperature fluctuations but also enhanced the minimum soil temperatures and declined the maximum soil temperatures. In summer, the mulching decreased soil temperature (by almost a half °C), and a vice versa effect was noted in the spring season (Suying et al. 2005). Similarly, maize straw applied in maize sown under the ridge-furrow system lowered the soil temperature (compared to control and black plastic mulch as well), which helped to improve the soil water availability through reduced evaporation (Li et al. 2013). The grain yield of maize was increased by 15% in the straw-mulched fields compared to control (Li et al. 2013). Maize straw was better in regulating the soil temperature than the gravel mulch (Jun et al. 2014). Mulch of wheat straw lowered the soil temperature as well as temperature fluctuations compared to control in green gram and maize during the summer season (Gaur and Mukherjee 1980). Rice straw mulch applied in *Brassica napus* L. decreased the soil temperature (~0.5–1.3 °C), and the impact was great during the start and middle of the crop growing season (Sarkar et al. 2007). Occasionally, negative impacts of straw mulch on the crop have also been observed. For example, mulch of oat could reduce the soil temperature in maize cultivated in Canada, and subsequently some damage to crop development was noted (Fortin and Pierce 1991).

Table 1 Impact of different types of mulches on soil temperature under the various crops

Type of mulch	Impact on soil temperature/crop growth	Crop	Reference
Coupled plastic and straw mulch	Decrease (1.26–1.31 °C) in temperature	Maize and wheat (intercropping)	Yin et al. (2016)
Straw mulch	Decrease in soil temperature fluctuations	Alfalfa	Jun et al. (2014)
	Decrease in soil temperature	Maize	Li et al. (2012)
	Decrease in soil temperature	Coffee	Gurnah and Mutea (1982)
Black plastic mulch	Increase in average, minimum and maximum temperatures	Cucumber	Torres-Olivar et al. (2018)
	Increase (1.2–2.3 °C) in soil temperature	Maize	Xiukang et al. (2015)
	Increase in soil temperature	Maize	Li et al. (2012)
	No impact on fluctuations of soil temperature	Vineyard	Pinamonti (1998)
	Regulation of soil temperature	Tomato	Teasdale and Abdul-Baki (1995)
	Increase in soil temperature	Coffee	Gurnah and Mutea (1982)
Transparent plastic mulch	Increase in soil temperature	Coffee	Gurnah and Mutea (1982)
	A 2–9 °C increase in soil temperature	Potato	Wang et al. (2009)
	Rise in soil temperature (2.5–3.2 °C), early crop emergence (8.1–11.7 days), and longer (0.7–15.0 days) growing season	Potato	Zhao et al. (2012)
	Increase in surface soil temperature by 1.0–2.5 °C/1.6–3.4 °C day/night and 0.4–1.3 °C decrease in diurnal temperature amplitude	Maize	Zhang et al. (2019)
Degradable film as mulch	Increase in soil temperature	Maize	Li et al. (2012)
Compost mulch	Decrease in temperature fluctuations	Vineyard	Pinamonti (1998)
Silver/black plastic mulch	Increase in average, minimum, and maximum temperatures	Cucumber	Torres-Olivar et al. (2018)

Conclusions

The discussion in this chapter indicated that other than raising or lowering the soil temperature, the most important impact of mulches has been the minimization of difference between the highest and lowest temperature during a day. In other words, the mulches help to minimize the temperature fluctuations. The reduced temperature swung is likely to cause positive impacts on crop plants and reduce the damages caused by the temperature extremes. Plastic mulches (mostly the black plastic) raise the soil temperature, and the straw mulches usually lower the soil temperature. This evidence provides an excellent opportunity to select the mulch type keeping in view the season, the soil and environmental conditions, and the crop. Black plastic mulch may be used in the low-temperature environments and in the crops that require warm soil conditions for seed germination and establishment of the crop stand. Similarly, the straw mulches may be utilized where the soil environment is hot and dry, and the crop requires a cool microclimate for its optimum growth and development. Care is required to avoid overheating damages caused to plant tissues (and probably to other biological communities in the soil as well) by the plastic mulches. Similarly, the straw mulch may decrease the soil temperatures lower than the optimum or the one required for root and crop growth, nutrient uptake, and other plant activities. Such damages should also be avoided by carefully selecting the time of mulch application. Optimization of the mulches for their type, time of application, and quantity may help to harness maximum of their benefits.

References

Decoteau, D.R., Kasperbauer, M.J. and Hunt, P.G., 1989. Mulch surface color affects yield of fresh-market tomatoes. *Journal Of The American Society For Horticultural Science*, *114*(2), 216–219.

Farooq, M., Aziz, T., Hussain, M., Rehman, H., Jabran, K. and Khan, M.B., 2008. Glycinebetaine improves chilling tolerance in hybrid maize. *Journal of Agronomy and Crop Science*, *194*(2), 152–160.

Farooq, M., Hussain, M., Nawaz, A., Lee, D.J., Alghamdi, S.S. and Siddique, K.H.M, 2017. Seed priming improves chilling tolerance in chickpea by modulating germination metabolism, trehalose accumulation and carbon assimilation. *Plant Physiology and Biochemistry*, *111*, 274–283.

Fortin, M.C. and Pierce, F.J., 1991. Timing and nature of mulch retardation of corn vegetative development. *Agronomy Journal*, *83*(1), 258–263.

Fortnum, B.A., Decoteau, D.R. and Kasperbauer, M.J., 1997. Colored mulches affect yield of fresh-market tomato infected with Meloidogyne incognita. *Journal of Nematology*, *29*(4), 538.

Gaur, A.C. and Mukherjee, D., 1980. Recycling of organic matter through mulch in relation to chemical and microbiological properties of soil and crop yields. *Plant and Soil*, *56*(2), 273–281.

Genyun, W.S.D., 1991. A study on the mechanism of soil temperature increasing under plastic mulch. *Scientia Agricultura Sinica*, *3*, 010.

Gurnah, A.M. and Mutea, J., 1982. Effects of mulches on soil temperatures under Arabica coffee at Kabete, Kenya. *Agricultural Meteorology*, *25*, 237–244.

Jun, F., Yu, G., Quanjiu, W., Malhi, S.S. and Yangyang, L., 2014. Mulching effects on water storage in soil and its depletion by alfalfa in the Loess Plateau of northwestern China. *Agricultural Water Management*, *138*, 10–16.

Kasperbauer, M.J., 2000. Strawberry yield over red versus black plastic mulch. *Crop Science*, *40*(1), 171–174.

Kumarathunge, D.P., Medlyn, B.E., Drake, J.E., Tjoelker, M.G., Aspinwall, M.J., Battaglia, M., Cano, F.J., Kelsey, C.R., Cavaleri, M.A., Cernusak, L.A., Chambers, J.Q. et al. 2019. Acclimation and adaptation components of the temperature dependence of plant photosynthesis at the global scale. *New Phytologist.*, *222*, 768–784.

Li, R., Hou, X., Jia, Z., Han, Q. and Yang, B., 2012. Effects of rainfall harvesting and mulching technologies on soil water, temperature, and maize yield in Loess Plateau region of China. *Soil Research*, *50*(2), 105–113.

Li, R., Hou, X., Jia, Z., Han, Q., Ren, X. and Yang, B., 2013. Effects on soil temperature, moisture, and maize yield of cultivation with ridge and furrow mulching in the rainfed area of the Loess Plateau, China. *Agricultural Water Management*, *116*, 101–109.

Nolan, C., Overpeck, J.T., Allen, J.R., Anderson, P.M., Betancourt, J.L., Binney, H.A., Brewer, S., Bush, M.B., Chase, B.M., Cheddadi, R. and Djamali, M., 2018. Past and future global transformation of terrestrial ecosystems under climate change. *Science*, *361*(6405), 920–923.

Pinamonti, F., 1998. Compost mulch effects on soil fertility, nutritional status and performance of grapevine. *Nutrient Cycling in Agroecosystems*, *51*(3), 239–248.

Rosbakh, S. and Poschlod, P., 2015. Initial temperature of seed germination as related to species occurrence along a temperature gradient. *Functional Ecology*, *29*(1), 5–14.

Ruiz, J.M., Hernandez, J., Castilla, N., Romero, L., 1999. Potato performance in response to different mulches. 1. Nitrogen metabolism and yield. *Journal of Agricultural and Food Chemistry*, *47* (7), 2660–2665.

Sarkar, S., Paramanick, M., Goswami, S.B., 2007. Soil temperature, water use and yield of yellow sarson (*Brassica napus* L. var. glauca) in relation to tillage intensity and mulch management under rainfed lowland ecosystem in eastern India. *Soil and Tillage Research*, *93*, 94–101.

Suying, C., Xiying, Z., Dong, P. and Hongyong, S., 2005. Effects of corn straw mulching on soil temperature and soil evaporation of winter wheat field. *Transactions of The Chinese Society of Agricultural Engineering*, *10*, 038.

Szymańska, R., Ślesak, I., Orzechowska, A. and Kruk, J., 2017. Physiological and biochemical responses to high light and temperature stress in plants. *Environmental and Experimental Botany*, *139*, 165–177.

Teasdale, J.R. and Abdul-Baki, A.A., 1995. Soil temperature and tomato growth associated with black polyethylene and hairy vetch mulches. *Journal of the American Society for Horticultural Science*, *120*(5), 848–853.

Torres-Olivar, V., Ibarra-Jiménez, L., Cárdenas-Flores, A., Lira-Saldivar, R.H., Valenzuela-Soto, J.H. and Castillo-Campohermoso, M.A., 2018. Changes induced by plastic film mulches on soil temperature and their relevance in growth and fruit yield of pickling cucumber. *Acta Agriculturae Scandinavica, Section B—Soil & Plant Science*, *68*(2), 97–103.

Wang, F.X., Feng, S.Y., Hou, X.Y., Kang, S.Z. and Han, J.J., 2009. Potato growth with and without plastic mulch in two typical regions of Northern China. *Field Crops Research*, *110*(2), 123–129.

Wang, W.H., Wang, Q.J. and Liu, J.J., 2011. Analysis of temperature changes under film during cotton seedling stage in south Xinjiang. *Agricultural Research in the Arid Areas*, *1*, 024.

Xiukang, W., Zhanbin, L. and Yingying, X., 2015. Effects of mulching and nitrogen on soil temperature, water content, nitrate-N content and maize yield in the Loess Plateau of China. *Agricultural Water Management*, *161*, 53–64.

Yin, W., Feng, F., Zhao, C., Yu, A., Hu, F., Chai, Q., Gan, Y. and Guo, Y., 2016. Integrated double mulching practices optimizes soil temperature and improves soil water utilization in arid environments. *International Journal of Biometeorology*, *60*(9), 1423–1437.

Zhang, X., Zhao, J., Yang, L., Kamran, M., Xue, X., Dong, Z., Jia, Z. and Han, Q., 2019. Ridge-furrow mulching system regulates diurnal temperature amplitude and wetting-drying alternation behavior in soil to promote maize growth and water use in a semiarid region. *Field Crops Research*, *233*, 121–130.

Zhao, H., Xiong, Y.C., Li, F.M., Wang, R.Y., Qiang, S.C., Yao, T.F. and Mo, F., 2012. Plastic film mulch for half growing-season maximized WUE and yield of potato via moisture-temperature improvement in a semi-arid agroecosystem. *Agricultural Water Management*, *104*, 68–78.

Zhu, J.K., 2016. Abiotic stress signaling and responses in plants. *Cell*, *167*(2), 313–324.

Mulches for Nutrient Addition to Soil

Introduction

Plants require nutrients for their optimum growth, development, and productivity. Nutrient concentrations in the rhizosphere are impacted by fertilizer application, soil-plant interactions, application of mulches or plant residue, microbial activities, and other factors. In some parts of the world, it is a common practice to incorporate plant residues, while in most parts of the world, the remains of crops are either burned, fed to animals, or used for other purposes (Erenstein 2002). Crop residues from the previous crops, if used as mulch, do bring several benefits to the soil including addition of organic matter and nutrients such nitrogen, potassium, carbon, phosphorus, etc. (Akhtar et al. 2018). Mulches possess a potential of improving the nutrients status and the other soil chemical and physical properties. Mulches of organic origin are a natural renewable source that can help in building the soil fertility (Qu et al. 2019). The role of mulches in soil carbon sequestration has also been mentioned previously (Lal 2004).

Several kinds of organic mulches are available; these may include (but not limited to) wood chips, tree cuttings, human waste, yard waste, sawdust, crop residues, poultry manure compost, compost of fruit remains, etc. (Ibeawuchi et al. 2015). This is important to select the right kind of mulch keeping in view the soil environment where these are being added. Mulches containing plant remains could be categorized as low or high quality based on their lignin concentration and C/N ratio. The ones with low lignin and C/N ratio were considered as high-quality residue, while the one with high lignin and C/N ratio were low-quality residue (Tian et al. 1993). The low-quality residue could improve the soil properties through mulching effect, while high-quality residue could support the crop by adding nutrients to soil (Tian et al. 1993).

Although mulching (permanent residue retention) brings about several benefits such as better soil quality, some negative impacts (such as decrease in air circulation

K. Jabran, *Role of Mulching in Pest Management and Agricultural Sustainability*, SpringerBriefs in Plant Science, https://doi.org/10.1007/978-3-030-22301-4_7

around the plant roots, high emission of nitrous oxide, increase in denitrification, etc.) may also be expected. Another aspect is decrease in soil organic matter after the mulch application. Mulch raises the soil temperature that can increase the decomposition of organic matter (Steinmetz et al. 2016). However, this kind of impact may be rarely noted.

This chapter is aimed at explaining the benefits of adding straw mulches to the soil particularly in the perspective of improving soil properties and addition of nutrients to the soil. As most part of crop residue is either lost or used for various purposes rather than incorporation into the soil, it is important to highlight the benefits that can be gained by using the crop straw as mulch in the next-season crop.

Organic Matter and Crop Residue Mulches

Soil organic matter concentration is among the most important indicators of soil fertility and soil fertility is mainly dependent on the extent of organic matter in the soil (Loveland and Webb 2003). Organic matter being the most dynamic fraction of soil has its specific role in improving soil structure, water holding capacity of the soil, nutrient retention capability of the soil, biological nitrogen fixation, soil aeration, root penetration, nutrient availability to plants, soil conservation by providing sticking power to soil separates, imparting suitable tilth to soil, supporting microflora and microfauna in the soil, and maintaining buffering and exchange capacity of the soil (Lal 2009; Pan et al. 2009).

It is imperative to maintain high reserves of organic matter in the soil (Lehmann and Kleber 2015). However, intensive agriculture in response to soaring food needs is a reason for decline in soil organic matter. Further, the buildup of soil organic matter under climatic conditions characterized with high temperature, low rainfall, and alkaline pH is highly difficult due to high decomposition rates (Kalbitz et al. 2000; Conant et al. 2011). Plant residues may be used as mulches to maintain organic matter contents in the soils. All nutrients removed from the soil by crop plants are stored in the crop herbage and grains along with the sun energy embodied in the form of different compounds. Hence, the crop residues, farm waste, and animal dung return the nutrients to soil when used as mulches. Soil organisms such as termites and earthworms start decomposing plant residues to convert them to tiny pieces when added to soil (Six et al. 2004). The residues, now in smaller size, are further subjected to action of other microorganisms to degrade micromolecules as starches, amino acids, proteins, sugars, etc., while the macromolecules as fats, waxes, lignin, etc., are first converted to simpler compounds for further breakdown (Lützow et al. 2006). Lignin-containing residues require more time to decompose. Microorganisms release certain enzymes to accomplish the decomposition process. Finally, the residues are converted to CO_2 which is utilized by soil microbes as energy source and become their body part and humus, the stable organic matter, which becomes lasting part of soil (Ajwa and Tabatabai 1994).

Stubble and Straw Mulches for Adding Nutrients to Soil

Both the stubble and straw are important organic mulches that add organic matter and nutrients to the soil (Table 1). Crop remains (stubble or straw) have been utilized as mulch in various ways depending on the prevailing cropping systems and other socio-economic factors. For example, cropping systems that retain the crop residue as mulch and seeded as no-till have been proposed as alternative to the cropping systems that comprise of conventional tillage and no retention of crop residue. The objective is to protect the soil from wind and water erosion, enhance carbon sequestration, increase biological activities in upper soil layer, improve physical properties of soil, and add nutrients to the soil (Blevins et al. 2018). No-till mulched cropping systems greatly increase the soil carbon storage and improve the soil health (Corbeels et al. 2006).

Recent literature clearly indicates that various organic mulches add significant amounts of nutrients to the soil. For example, a 3-year study from China tested the *Imperata cylindrical* var. major mulch (0, 10, 20 and 30 t/ha) for nutrient addition to soil and improved the growth of poplar (*Populus deltoides* × *P. nigra* cv. Zhonglinmeihe) (Fang et al. 2008). The mulch enhanced the concentration of available nutrients (potassium, nitrogen, and phosphorus) and increased the poplar height, breadth, and dry matter production over control. The higher rates (20 and 30 t/ha) of mulch were more effective than the lower rate (10 t/ha); however, the two higher rates (20 and 30 t/ha) mostly provided similar results. A mulch rate of 20 t/ha in the newly planted poplar stem cuttings may be recommended particularly in the poor soils (Fang et al. 2008). In the poplar plantations on the degraded soils,

Table 1 Effect of different kinds of mulches on the soil properties or nutrient additions to the soil

Type of mulch	Impact on soil nutrients/properties	References
Compost of chicken	Increase in soil pH	Cooper and Warman (1997)
Leaves of maple (*Acer* spp.)	Increase in total soil nitrogen and carbon and infiltration rate	Acosta-Martinez et al. (1999)
Wheat straw	A 30% increase in nitrogen potential availability	Tu et al. (2006)
Straw mulch	Increase in soil carbon and nitrogen	Dawes (2010)
Mulch of maize stalks	Increase in soil organic matter	Chen et al. (2014)
Plastic mulch	Increase in soil organic carbon	Zhang et al. (2017)
Wheat straw mulch	Increase in soil organic matter, organic carbon in the soil, and enzymatic activities in the soil and addition of nitrogen and phosphorus to soil along with an increased availability of these nutrients	Akhtar et al. (2018)
Compost of green waste	Increase in soil porosity and fertility	Qu et al. (2019)

mulches from *Quercus fabri* Hance, *I. cylindrica*, *Pteridium aquilinum*, and *Coriaria nepalensis* Wallich enhanced the annually available nitrogen (~30–40%) and the mineralization of nitrogen (22–30%) (Fang et al. 2011). These mulches also improved the long-term availability of several soil nutrients and increased the poplar height, biomass, and breadth over the control (Fang et al. 2011). In another 5-year study, increase in available phosphorus and nitrate-nitrogen was noted with the application of straw mulch from maize (Jun et al. 2014).

Mulches may also negatively impact the dynamics of nutrients in the soil or the uptake of nutrients by the crop plants. For example, wheat straw mulch decreased the nutrient uptake and yield of (non-flooded) rice in a rice-wheat system but increased the soil nitrogen-phosphorus-potassium balance (Liu et al. 2003). In another study, rice straw was used as mulch in winter oilseed rape (*Brassica napus* L.), and this increased the uptake of nitrogen as well as crop biomass, but the mulch layer also resulted in nitrogen loss through ammonia volatilization (Su et al. 2014). This emphasizes the need for fertilizer application methods other than topdressing in case of straw application.

Mulches from Woody Plants

Parts of woody plants or residues from the wood plants may act as a source of mulching material to add nutrients to soil and improve its properties (Li et al. 2018). For example, cuttings of *Leucaena leucocephala* added high concentrations of nutrients [phosphorus = 84%, nitrogen = 96%, magnesium = 50%] to soil and caused the highest increase in maize yield over control and other treatments in the experiment (Tian et al. 1993). Similarly, addition of *Acioa barteri* residue to maize had also increased the plant nutrient uptake [calcium = 92%, potassium = 59%] and increased the crop grain yield (Tian et al. 1993). In another study, *Leucaena* was used as mulch in the coffee plantations and found to increase organic carbon and nitrogen availability (Youkhana and Idol 2009). From this brief discussion, it can be concluded that woody plants or their residues possess a potential to be used as mulch in some kinds of agricultural production and subsequently add nutrients to soil and improve it properties.

Compost Mulch for Adding Nutrients to Soil

Composts are an important source of plant nutrients and have been particularly useful in the organic crop production systems. A wide variety of materials (such as plant trash, fruit remains, wood ash, paper sludge, pulp, etc.) can be used to produce compost for use as a mulch in the crop production (Campbell et al. 1995; Qu et al. 2019). Composts have been used as mulch not only for adding nutrients to soil but also for other purposes such as improving soil properties, soil conservation, increasing soil water holding capacity, and suppressing weeds or others of the soilborne plant pests. Generally, mulching of compost adds higher quantities of organic matter and other nutrients to the soil compared to other mulch types. For example, compost mulch was compared with wood mulch and no mulch application for their impact on soil properties (Tiquia et al. 2002). Compost caused a significant increase in the soil organic matter and concentrations of calcium, magnesium, and potassium (Tiquia et al. 2002). Similarly, a comparison between sawdust and compost mulches indicated that compost not only added higher quantities of organic matter (11 g/kg) to soil but also supplemented several of other nutrients to soil (Larco et al. 2013). Soil applied with a combination of sawdust and compost mulches had higher concentrations of zinc, boron, phosphorus, potassium, and nitrogen than the ones applied either with sawdust or compost alone (Larco et al. 2013).

Other Mulches for Adding Nutrients to Soil

Plastic mulch takes place among the non-organic types of mulches that do not add nutrients to the soil. Nevertheless, the plastic mulching can positively impact the nutrient cycling in the soil. Several of the studies indicate that nitrogen losses are decreased under plastic mulching, and plant nitrogen uptake is likely to increase. For instance, an improved C/N ratio was noted under plastic mulching on a ridge-furrow system (Mo et al. 2017). The same system increased the uptake and productivity of nitrogen in the maize crop (Li et al. 2016). Plastic mulch also increased the nutrient uptake in tomatoes along with increase in its root growth, flowering, plant dry matter, and branching (Wien et al. 1993). Plastic mulch could also increase the nutrient uptake in rice (grown under non-flooded environment), but the soil balance of nitrogen-phosphorus-potassium was lowered by the end of growing season (Liu et al. 2003). Application of plastic mulch was also found to decrease the leaching of nitrogen under drip irrigation and fertigation (Filipović et al. 2016).

Application of an organic mulch for a long period may bring benefits more than nutrient addition to soil. For example, almond shell mulch applied in avocado orchard formed a new organic layer over the soil after a period of 10 years (López et al. 2014). Phosphorus, nitrogen, and carbon contents were increased with the almond mulch. The lower lignin concentration in this mulch reduced the rate of organic matter decay that prolonged the nutrient availability to the avocado trees (López et al. 2014). Similarly, a combination of organic and non-organic mulches could also cause great improvement in the soil characteristics. For example, plastic mulch combined with poplar tree leaves caused a highest increase in soil organic carbon, humic acid, and cation exchange capacity compared with maize straw, sheep manure, or fodder grass (Hu et al. 2018).

Conclusions

In general, the application of organic mulches increases the concentration of organic matter and other nutrients in the soil. However, little is known about the impacts of mulches on nutrient cycling and dynamics in the soil. There is need for care about the nitrogen application as topdressing to straw, plastic, or other mulches (Wang et al. 2018). The nitrogen fertilizer applied this way may be lost; nitrogen fertilizer should be applied through special equipment and near to root surface. Dripping of nitrogen is likely to be beneficial. Plastic mulch does not directly add organic matter or nutrients to soil, but it increases the nitrogen availability and reduces its loss through leaching. It is also concluded that a combined application of more than one mulch may better improve the soil properties. Similarly, in some cases, there can be a possibility to combine fertilizer and mulch application to achieve their improved impact on soil properties and crop growth (Shi et al. 2009).

References

Acosta-Martinez, V., Reicher, Z., Bischoff, M. and Turco, R.F., 1999. The role of tree leaf mulch and nitrogen fertilizer on turfgrass soil quality. *Biology and Fertility of Soils*, 29(1), 55–61.

Ajwa, H.A. and Tabatabai, M.A., 1994. Decomposition of different organic materials in soils. *Biology and Fertility of Soils*, 18(3), 175–182.

Akhtar, K., Wang, W., Ren, G., Khan, A., Feng, Y. and Yang, G., 2018. Changes in soil enzymes, soil properties, and maize crop productivity under wheat straw mulching in Guanzhong, China. *Soil and Tillage Research*, 182, 94–102.

Blevins, R.L., Lal, R., Doran, J.W., Langdale, G.W. and Frye, W.W., 2018. Conservation tillage for erosion control and soil quality. In: Pierce, F.J. (ed.) *Advances in Soil and Water Conservation* (pp. 51–68). Routledge, Taylor & Francis Group, United States.

Campbell, A.G., Zhang, X. and Tripepi, R.R., 1995. Composting and evaluating a pulp and paper sludge for use as a soil amendment/mulch. *Compost Science & Utilization*, 3(1), 84–95.

Chen, Y., Wen, X., Sun, Y., Zhang, J., Wu, W. and Liao, Y., 2014. Mulching practices altered soil bacterial community structure and improved orchard productivity and apple quality after five growing seasons. *Scientia Horticulturae*, 172, 248–257.

Conant, R.T., Ryan, M.G., Ågren, G.I., Birge, H.E., Davidson, E.A., Eliasson, P.E., Evans, S.E., Frey, S.D., Giardina, C.P., Hopkins, F.M. and Hyvönen, R., 2011. Temperature and soil organic matter decomposition rates–synthesis of current knowledge and a way forward. *Global Change Biology*, 17(11), 3392–3404.

Cooper, J.M. and Warman, P.R., 1997. Effects of three fertility amendments on soil dehydrogenase activity, organic C and pH. *Canadian Journal of Soil Science*, 77(2), 281–283.

Corbeels, M., Scopel, E., Cardoso, A., Bernoux, M., Douzet, J.M. and Neto, M.S., 2006. Soil carbon storage potential of direct seeding mulch-based cropping systems in the Cerrados of Brazil. *Global Change Biology*, 12(9), 1773–1787.

Dawes, T.Z., 2010. Reestablishment of ecological functioning by mulching and termite invasion in a degraded soil in an Australian savanna. *Soil Biology and Biochemistry*, 42(10), 1825–1834.

Erenstein, O., 2002. Crop residue mulching in tropical and semi-tropical countries: An evaluation of residue availability and other technological implications. *Soil and Tillage Research*, 67(2), 115–133.

Fang, S., Xie, B. and Liu, J., 2008. Soil nutrient availability, poplar growth and biomass production on degraded agricultural soil under fresh grass mulch. *Forest Ecology and Management*, 255(5–6), 1802–1809.

Fang, S., Xie, B., Liu, D. and Liu, J., 2011. Effects of mulching materials on nitrogen mineralization, nitrogen availability and poplar growth on degraded agricultural soil. *New Forests*, 41(2), 147–162.

Filipović, V., Romić, D., Romić, M., Borošić, J., Filipović, L., Mallmann, F.J.K. and Robinson, D.A., 2016. Plastic mulch and nitrogen fertigation in growing vegetables modify soil temperature, water and nitrate dynamics: Experimental results and a modeling study. *Agricultural Water Management*, 176, 100–110.

Hu, J., Wu, J., Qu, X. and Li, J., 2018. Effects of organic wastes on structural characterizations of humic acid in semiarid soil under plastic mulched drip irrigation. *Chemosphere*, 200, 313–321.

Ibeawuchi, I.I., Iwuanyanwu, U.P., Nze, E.O., Olejeme, O.C. and Ihejirika, G.O., 2015. Mulches and organic manures as renewable energy sources for sustainable farming. *Journal of Natural Sciences Research*, 5(2), 139–147.

Jun, F., Yu, G., Quanjiu, W., Malhi, S.S. and Yangyang, L., 2014. Mulching effects on water storage in soil and its depletion by alfalfa in the Loess Plateau of northwestern China. *Agricultural Water Management*, 138, 10–16.

Kalbitz, K., Solinger, S., Park, J.H., Michalzik, B. and Matzner, E., 2000. Controls on the dynamics of dissolved organic matter in soils: a review. *Soil Science*, 165(4), 277–304.

Lal, R., 2004. Soil carbon sequestration impacts on global climate change and food security. *Science*, 304(5677), 1623–1627.

Lal, R., 2009. Challenges and opportunities in soil organic matter research. *European Journal of Soil Science*, 60(2), 158–169.

Larco, H., Strik, B.C., Bryla, D.R. and Sullivan, D.M., 2013. Mulch and fertilizer management practices for organic production of highbush blueberry. II. Impact on plant and soil nutrients during establishment. *HortScience*, 48(12), 1484–1495.

Lehmann, J. and Kleber, M., 2015. The contentious nature of soil organic matter. *Nature*, 528(7580), 60–68.

Li, C., Wen, X., Wan, X., Liu, Y., Han, J., Liao, Y. and Wu, W., 2016. Towards the highly effective use of precipitation by ridge-furrow with plastic film mulching instead of relying on irrigation resources in a dry semi-humid area. *Field Crops Research*, 188, 62–73.

Li, Z., Schneider, R.L., Morreale, S.J., Xie, Y., Li, C. and Li, J., 2018. Woody organic amendments for retaining soil water, improving soil properties and enhancing plant growth in desertified soils of Ningxia, China. *Geoderma*, 310, 143–152.

Liu, X.J., Wang, J.C., Lu, S.H., Zhang, F.S., Zeng, X.Z., Ai, Y.W., Peng, S.B. and Christie, P., 2003. Effects of non-flooded mulching cultivation on crop yield, nutrient uptake and nutrient balance in rice–wheat cropping systems. *Field Crops Research*, 83(3), 297–311.

López, R., Burgos, P., Hermoso, J.M., Hormaza, J.I. and González-Fernández, J.J., 2014. Long term changes in soil properties and enzyme activities after almond shell mulching in avocado organic production. *Soil and Tillage Research*, 143, 155–163.

Loveland, P. and Webb, J., 2003. Is there a critical level of organic matter in the agricultural soils of temperate regions: a review. *Soil and Tillage research*, 70(1), 1–18.

Lützow, M.V., Kögel-Knabner, I., Ekschmitt, K., Matzner, E., Guggenberger, G., Marschner, B. and Flessa, H., 2006. Stabilization of organic matter in temperate soils: mechanisms and their relevance under different soil conditions–a review. *European Journal of Soil Science*, 57(4), 426–445.

Mo, F., Wang, J.Y., Zhou, H., Luo, C.L., Zhang, X.F., Li, X.Y., Li, F.M., Xiong, L.B., Kavagi, L., Nguluu, S.N. and Xiong, Y.C., 2017. Ridge-furrow plastic-mulching with balanced fertilization in rainfed maize (Zea mays L.): An adaptive management in east African Plateau. *Agricultural and Forest Meteorology*, 236, 100–112. 4.

Pan, G., Smith, P. and Pan, W., 2009. The role of soil organic matter in maintaining the productivity and yield stability of cereals in China. *Agriculture, Ecosystems & Environment, 129*(1–3), 344–348.

Qu, B., Liu, Y., Sun, X., Li, S., Wang, X., Xiong, K., Yun, B. and Zhang, H., 2019. Effect of various mulches on soil physico—Chemical properties and tree growth (*Sophora japonica*) in urban tree pits. *PLoS One, 14*(2), e0210777.

Six, J., Bossuyt, H., Degryze, S. and Denef, K., 2004. A history of research on the link between (micro) aggregates, soil biota, and soil organic matter dynamics. Soil and Tillage Research, 79(1), 7–31.

Shi, Z.H., Chen, L.D., Cai, C.F., Li, Z.X. and Liu, G.H., 2009. Effects of long-term fertilization and mulch on soil fertility in contour hedgerow systems: a case study on steeplands from the Three Gorges Area, China. *Nutrient Cycling in Agroecosystems, 84*(1), 39–48.

Steinmetz, Z., Wollmann, C., Schaefer, M., Buchmann, C., David, J., Tröger, J., Muñoz, K., Frör, O. and Schaumann, G.E., 2016. Plastic mulching in agriculture. Trading short-term agronomic benefits for long-term soil degradation?. *Science of the Total Environment, 550*, 690–705.

Su, W., Lu, J., Wang, W., Li, X., Ren, T., Cong, R., 2014. Influence of rice straw mulching on seed yield and nitrogen use efficiency of winter oilseed rape (*Brassica napus* L.) in intensive rice–oilseed rape cropping system. *Field Crops Research, 159*, 53–61.

Tian, G., Kang, B.T. and Brussaard, L., 1993. Mulching effect of plant residues with chemically contrasting compositions on maize growth and nutrients accumulation. *Plant and Soil, 153*(2), 179–187.

Tiquia, S.M., Lloyd, J., Herms, D.A., Hoitink, H.A. and Michel Jr, F.C., 2002. Effects of mulching and fertilization on soil nutrients, microbial activity and rhizosphere bacterial community structure determined by analysis of TRFLPs of PCR-amplified 16S rRNA genes. *Applied Soil Ecology, 21*(1), 31–48.

Tu, C., Ristaino, J.B., Hu, S., 2006. Soil microbial biomass and activity in organic tomato farming systems: Effects of organic inputs and straw mulching. *Soil Biology and Biochemistry, 38*, 247–255

Wang, X., Fan, J., Xing, Y., Xu, G., Wang, H., Deng, J., Wang, Y., Zhang, F., Li, P., and Li, Z., 2018. The Effects of mulch and nitrogen fertilizer on the soil environment of crop plants. *Advances in Agronomy, 153*, 121–174.

Wien, H.C., Minotti, P.L. and Grubinger, V.P., 1993. Polyethylene mulch stimulates early root growth and nutrient uptake of transplanted tomatoes. *Journal of the American Society for Horticultural Science, 118*(2), 207–211.

Youkhana, A. and Idol, T., 2009. Tree pruning mulch increases soil C and N in a shaded coffee agroecosystem in Hawaii. *Soil biology and Biochemistry, 41*(12), 2527–2534.

Zhang, F., Zhang, W., Li, M., Yang, Y. and Li, F.M., 2017. Does long-term plastic film mulching really decrease sequestration of organic carbon in soil in the Loess Plateau?. *European Journal of Agronomy, 89*, 53–60.

Index

© The Author(s), under exclusive licence to Springer Nature Switzerland AG 2019
K. Jabran, *Role of Mulching in Pest Management and Agricultural
Sustainability*, SpringerBriefs in Plant Science,
https://doi.org/10.1007/978-3-030-22301-4

Printed in the United States
By Bookmasters